MAP
PROJECTION
METHODS

Frederick Pearson, II

Computer Sciences Corporation

 Sigma Scientific, Inc.

Blacksburg, Virginia

1984

MAP PROJECTION METHODS

This book was printed using an NBI System 64 Word Processing System at Virginia Polytechnic Institute and State University. Bookcrafters, Inc., of Ann Arbor, Michigan, reproduced and bound this book.

LIBRARY OF CONGRESS CATOLOGING IN PUBLICATION DATA

Pearson, II, Frederick
Map Projection Methods
(The Sigma Series on Applied Mathematics for Science and Engineering)
Bibliography, index, 130 figures, 292 pages.
1. Cartography, 2. Geodesy, 3. Applied Mathematics
Library of Congress Card Catalog Number 83-51506
ISBN 0-915313-00-6

MAP
PROJECTION
METHODS

THE SIGMA SERIES IN APPLIED MATHEMATICS FOR SCIENCE AND ENGINEERING

Editor

John L. Junkins
Professor of Engineering Science and Mechanics
Virginia Polytechnic Institute and State University
Blacksburg, Virginia

Preface

In this text, map projection equations are derived for the direct transformation from latitude and longitude to cartesian map coordinates for the most important mapping schemes. The unifying principles of differential geometry are applied to produce equal area, conformal, and conventional projections. In many cases the inverse transformation from cartesian coordinates to latitude and longitude are also derived.

The derivations proceed from first principles to usable equations which produce results suitable for hand plotting, digital/analog plotting, or display on a CRT.

This work was inspired because of a serious lack of a unified and comprehensive text concerning map projection theory. Most of the material presented in this text is available in the literature but is scattered in many texts, reports, documents and papers. This text develops in one self-contained source, the major map projections useful to mathematicians, scientists, and engineers.

In chapter one, the problem is stated, and the reader is introduced to the terminology of the art of map projections. In chapter two, basic transformation theory is introduced, and then particularized to the transformation of points on a sphere or spheroid onto a developable surface. The criterion employed for the derivations throughout is to use the most simple and direct approach.

The model of the earth is considered in chapter three. The most recent numerical values for parameters to describe the figure of the earth are given, and tables incorporating these are included for meridian length, parallel length and the relation between geodetic and geocentric latitude.

Equal area, conformal, and conventional map projection equations are derived in chapters four, five, and six, respectively. The final plotting equations are given in a form suitable for hand or computer computation. Plotting tables for the mapping grid have been generated for the most important of the projections. Since the proof of all the derivations is a correct graticule of meridians and parallels, original figures of these have been produced. The plotting tables and the figures reflect the modern parameters for the earth.

The most rigorous criteria for the success of a map projection scheme, and the tool for selecting the most useful scheme for a specific application is obtained by considering the theory of distortions. This is done in chapter seven. A numerical method is introduced which permits a quantitative estimate of linear and angular distortions.

Finally, chapter eight gives current and recommended uses for the various map projections. It is intended to aid the maker or user of maps in his choice of the correct projection for his particular application.

My thanks go to Dr. Richard Anderle of the Naval Surface Weapons Center, and Dr. John Junkins of Virginia Polytechnic Institute and State University for review and encouragement. In addition, Dr. Junkins served as technical editor for this manuscript. My especial thanks go to Mary Ellen Pearson for typing a draft of this manuscript, Everett George for his assistance with illustrations, and to Vanessa McCoy for her expert preparation of the final manuscript.

I dedicate this text to my wife, Mary Ellen.

Contents

Figures

Tables

MAP PROJECTION METHODS

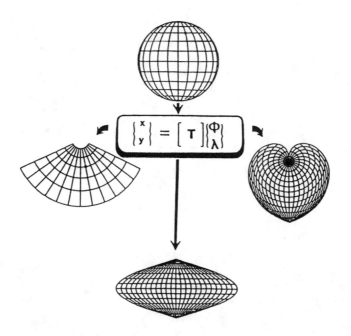

1

Introduction

Map projection is the orderly transfer of positions of places on the surface of the earth to corresponding points on a flat sheet of paper, a map. Since the surface of a sphere cannot be laid flat on a plane, without distortion, the process of transformation requires a degree of approximation and simplification. This first chapter lays the ground-work for this subject by detailing, in a qualitative way, the basic problem and introducing the nomenclature of maps. Succeeding chapters consider the mathematical techniques and the simplifications required to obtain manageable solutions [9].*

All projections introduce distortions in the map. The types of distortion are considered in terms of length, angle, and area. This first chapter discusses the qualitative aspects of the problem, while Chapter 7 deals with it quantitatively.

The coordinate systems useful in locating positions on the earth, and on the map are summarized. The concept of scale factor to reduce earth sized lengths to map sized lengths is discussed.

Map projections may be classified in a number of ways. The principle one is by the features preserved from distortion by the mapping technique. Other methods of classification depend on the plotting surface employed, the method of contact of this surface with the earth, the orientation of the plotting surface with respect to the direction of the earth's polar axis and whether the plotting surface is tangent or secant to the earth. Finally, maps can be

---

*Numbers in the brackets refer to the bibliography.

---

classified according to whether or not a map can be drawn by purely graphical means.

The convention for azimuth used in this volume is also introduced, as is the concept of the constant of a cone.

1.1 Introduction to the Problem

Map projection requires the transformation of positions from a curved surface, the earth, onto a plane surface, the map, in an orderly fashion. The problem occurs because of the difference in the surfaces involved.

The model of the earth is either a sphere or a spheroid (Chapter 3). These curved surfaces have two finite radii of curvature. The map is a plane surface, and a plane is characterized by two infinite radii of curvature. As is shown in Chapter 2, it is impossible to transform from a surface of two finite radii of curvature to a surface of two infinite radii of curvature without introducing some distortion. The sphere and the spheroid are called nondevelopable surfaces. This refers to the inability of these surfaces to be developed (i.e., laid flat) onto a plane in a distortion free manner [8].

Intermediate between the nondevelopable sphere and the spheroid, and the plane are surfaces with one finite and one infinite radius of curvature. The examples of this type of surface are the cylinder and the cone. These surfaces are called developable. Both the cylinder and the cone can be cut, and then developed (essentially unrolled along the finite radius of curvature) to form a plane. This development introduces no distortion, and thus, these figures may be used as intermediate plotting surfaces between the sphere and spheroid, and the plane. However, in any transformation from the sphere or spheroid to the developable surface some damage has already been done. The transformation from the nondevelopable to the developable surface invariably introduces some degree of distortion.

Consider the following properties of an ideal map [8]:

(1) Areas on the map maintain correct proportion to areas on the earth.

---

(2) Distances on the map would remain in true scale.

(3) Directions and angles on the map would remain true.

(4) Shapes on the map would be the same as on the earth.

The impossibility of distortion free transformation from the nondevelopable surface to the plane prevents the realization of the ideal. The best a cartographer can hope for is a realization of one or two of these features over the entire map. The other ideal properties are subject to distortion, but hpefully to a controlled extent.

The projections of Chapters 4, 5, and 6 are motivated directly by these ideal properties. In each of these projections, some of the desired features are maintained. The distortion in the other features is then tolerable.

1.2 Distortions [24]

Distortion is the villain of the piece. Distortion in maps may be in area, length, angle, or shape. Distortion is an untrue representation of these qualities of area, length, angles, or shape caused by following a specific map projection method.

Distortion in area is shown in Figure 1.2.1(a). While shape is maintained, the area on the map may be enlarlged or diminished. Distortion in length is common, and Figure 1.2.1(b) is an illustration. Often, while the cartographer is able to maintain true length in one direction, he cannot do so in a second direction. Angular distortion is also prevalent. Thus, angles on a map will not necessarily be the same as their counterparts on the earth. Thus, azimuths on the map, α' will not coincide with true azimuths α on the earth. This is shown in Figure 1.2.1(c). Distortion in shape can occur in a number of ways. One is a general change of shape of the figure. A second is a shearing type of effect. Figure 1.2.1(d) demonstrates both of these changes.

An actual map has combinations of these distortions. The numberical theory of distortion is presented in Chapter 7.

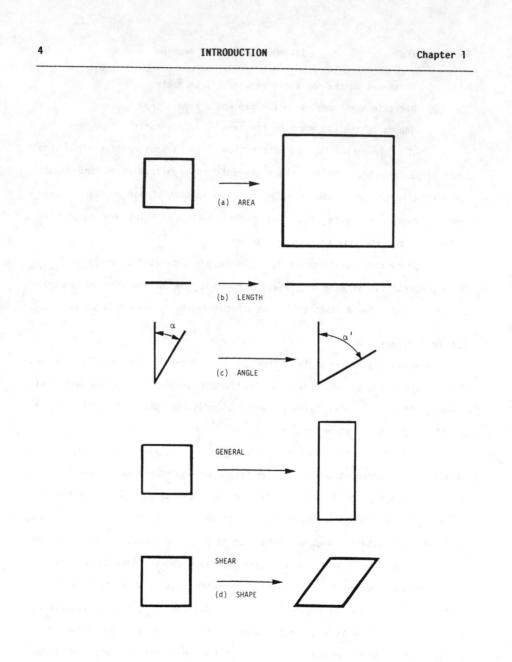

Figure 1.2.1 Distortion Effects

Now that we are acquainted with the problem, the next sections in this chapter introduce some of the terms needed for the study of map projections before entering the mathematics of Chapter 2.

1.3 Coordinate Systems [18]

Coordinate systems are necessary for both the earth and maps for the orderly locations of points. Two types of coordinate systems will be considered for the earth. They are a cartesian system, and an angular system. For maps, the most convenient system is cartesian.

The terrestrial coordinate system is demonstrated in Figure 1.3.1. The origin, 0, of the system is at the center of the earth. The x- and y-axis form the equatorial plane. The curve on the earth formed by the intersection of this plane with the earth's surface is the equator. The positive x-axis intersects the curve AGN. The curve AGN is a plane curve, which is called the Greenwich meridian. The positive z-axis coincides with the nominal axis of rotation of the earth, and points in the direction of the north pole, N. The z-axis completes a right-handed coordinate system.

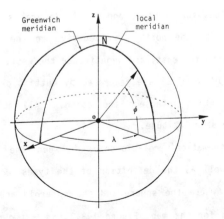

Figure 1.3.1   Terrestrial Coordinate System

Any point, P, on the surface of the earth can be located by the coordinates x, y, and z. However, since any point is constrained to lie on the surface, the three coordinates are not all independent. They are related by the equation of the surface (Chapter 3). Thus, there are only two independent coordinates, or two geometric degrees of freedom.

Instead of using two arbitrarily chosen members of the set x, y, and z as the independent coordinates, it is more convenient to use two independent angular (spherical) coordinates: latitude and longitude.

A meridian is a curve formed by the intersection of a ficticious plane containing the z-axis and the surface of the earth. The Greenwich meridian has already been mentioned. There is an infinity of meridians, depending on the orientation of the cutting plane.

The use of latitude and longitude locates a point on a meridian (latitude) and then locates (longitude) the meridian with respect to the Greenwich meridian. Latitude is the angular measure defining the position of point P on the meridian BPN. Latitude is denoted by ϕ. The position of the meridian that contains P is defined by the longitude, λ. The longitude is the angle AOB, measured in the equatorial plane, from the Greenwich meridian.

The conventions for latitude and longitude are as follows: latitude is measured positive to the north, and negative to the south. Longitude is measured positive to the east, and negative to the west.

The circles of parallel are generated by cutting plane 5 parallel to the equatorial plane which intersect the earth. All points on a circle of parallel have the same latitude.

The mathematical relationships between the polar and cartesian coordinates, as well as the definition of the types of latitude is deferred until Chapter 3, where the sphere and the spheroid are discussed. The coordinate system for the map, Figure 1.3.2, is a two dimensional cartesian system. The plus x-axis is toward the east, and the plus y-axis is toward the

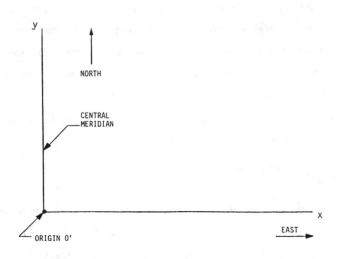

Figure 1.3.2   Map Coordinate System

north. The origin, 0', of the system will depend on the scheme of projection
to be developed in Chapters 4, 5, and 6. In most cases there will be some
straight, arbitrarily chosen central meridian which serves as the ordinate of
the projection.

The object of map projection is to transform from the terrestial angular
system to the map Cartesian system. Chapters 4, 5, and 6 provide the methods
for these transformations.

1.4 Scale [8]

The scale of the map is the ratio of the distance on the map to the
corresponding distance on the earth, or d_m/d_e. If the distance on both the
map and the earth have the same units, then the scale is a dimensionless
quantity. Scale is another aspect of the orderly transformation from earth
measurement to map measurement.

The presence of distortion requires the definition of two types of scale: the principal scale and the local scale. The principal scale is based on a meridian or parallel which is uniformly true scale for the entire map. It is the scale used for shrinking the spehrical or spheroidal surface of the earth to the plane of the paper. At other places on the map, where distortions are present, the scale will be different from true scale. This local scale will be larger or smaller than the principal scale, depending on the mechanism of the distortion. With d as a distance on a map,

$$S_{LOCAL} = \frac{d\ DISTORTED}{d\ TRUE\ LENGTH} \qquad\qquad (1.4.1)$$

For example, given that the principal scale of a map at a true length parallel is 1/10,000 (or 1" $\equiv$ 10,000'), and that the local scale at a second parallel is 0.9900. The relation between the length of the second parallel on the map, and its equivalent on the earth is:

(1)(0.9900)" $\equiv$ 10,000'

or

1" $\equiv$ 10,101'

The local scale, as a function of distortion, and the principal scale may be quoted on the legend of a map. A more useful means is a graphical scale drawn in the map legend, and specified for the latitudes and longitudes where it applies. As an example, for the Mercator projection (Chapter 5), a set of scales can be drawn as a function of latitude, which will ensure the correct basic distances for measuring.

The terms large scale versus small scale come from consideration of the fraction d_m/d_e. A scale of 1/10,000 is a large scale, and 1/1,000,000 is a small scale. A plan, or a map showing buildings, cultural features, or boundaries is usually 1/10,000 or larger. A topographic map, which gives roads, railroads, towns, and contour lines, and other details has a scale between 1/10,000 and 1/1,000,000. Maps of a scale smaller than 1/1,000,000 are atlas maps. These maps delineate countries, continents, and oceans.

The scale factor, S, or principal scale, is used in the plotting equations of Chapters 4, 5, and 6, and is equal to d_m/d_e: $S = d_m/d_e$.   The local scale of Chapter 7 is denoted by m.

1.5 Classification by Feature Preserved [24]

Any classification is an orderly means of organizing information. The classifications are based on a characteristic that sets one projection appart from others.

Maps may be classified by the feature rescued from distortion, or by the agreement that some distortion will simply be tolerated. This system of classification divides maps into three categories: equal area, conformal and conventional.

The equal area projection preserves the ratio of areas on the earth and on the map as a constant. Any part of the map bears the same relation to the area on the earth it represents that the whole map bears to the total earth area represented. Any quadrangular shaped section of the map formed by a grid of meridians and parallels will be equal in area to any other quadrangular area of the same map that represents an equal area of the earth. Angles usually suffer significant distortion. A contraction of meridians will have to be offset by a lengthening of parallels, or vice versa, but the enclosed area will remain the same. This concept is illustrated in Figure 1.5.1(a). All of the quadrilaterals have the same area.

A conformal projection is one in which the shape of any small surface of the map is preserved in its original form. Care must be used in applying this concept, since it is true only locally, and cannot be extended over large surface areas. The true condition for a conformal map is that the scale at any point is the same in all directions. The scale will change from point to point, but it will be independent of the azimuth at all points. The scale

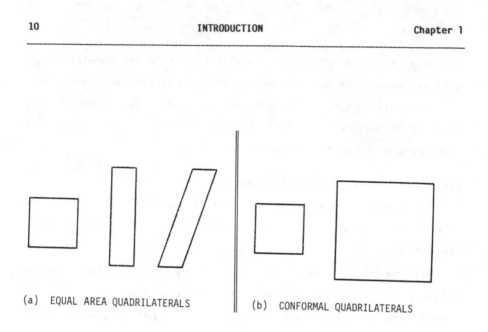

(a) EQUAL AREA QUADRILATERALS   (b)   CONFORMAL QUADRILATERALS

Figure 1.5.1 Quadrilateral Representation

will be the same in all directions from a point if two directions at right angles on the earth are mapped into two directions that are also at right angles to each other.   The meridians and parallels of the earth intersect at right angles; a conformal projection preserves this quality on the map and, thus, angular distortion is locally zero.   Conformal quadrilaterals are shown in Figure 1.5.1(b).   Another term used in referring to conformal projections is orthomorphic, or same form.

Conventional projections are all those which are neither equal area nor conformal.   This is not meant as a disparaging term.   Many of the conventional maps are of great utility.   In the Gnomonic projection, the feature preserved is that great circles become straight lines.   In the Azimuthal equidistant projection the distance and azimuth from the origin to any other point on the map is true.   The Polyconic and van der Grinten projections have seen considerable service as road maps. All that is implied by the term conventional is that the cartographer has been willing to sacrifice the features of equal area of conformality in order to retain some other desired feature, or to obtain a simple, utilitarian algorithm for the projection.

1.6 Classification by Projection Surface [24]

Only three projection surfaces will be considered - the plane, the cone, and the cylinder. Virtually all projections in use today are accomplished through these, or modifications of these. It can be argued that all projection surfaces are conical, since the plane and the cylinder can be considered as the two limiting cases of the cone. This mathematical concept is used in some cases, but the three surfaces are considered as distinct, in many others. Figure 1.6.1 shows each of these surfaces in relation to the sphere.

The planar projection surface can be used for a direct transformation from the earth. The projections which result are called azimuthal (Figure 1.6.1(a)). Other names in use are zenithal, or planar projections.

Conical projections result when a cone is used as an intermediate plotting surface. The cone is then developed into a plane to obtain the map (Figure 1.6.1(b)).

Cylindrical projections are obtained when a cylinder is used as the intermediate plotting surface (Figure 1.6.1(c)). As with the cone, the cylinder can then be developed into a plane.

In Figure 1.6.1, a representative position on the earth, P, is shown transformed into a position on the projection surface, P' for each of the projection surfaces, Chapters 4, 5, and 6 explain the methods that affect such transformations, and produce useful maps.

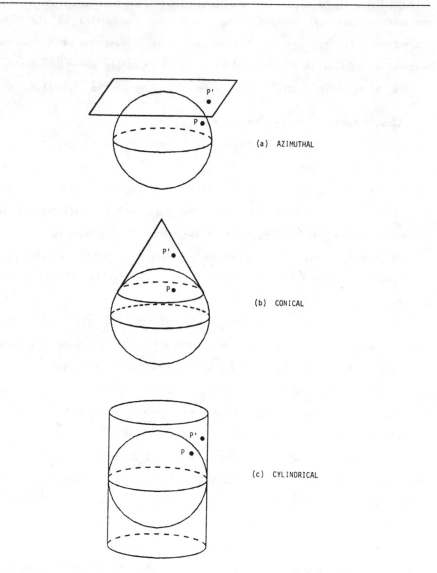

(a) AZIMUTHAL

(b) CONICAL

(c) CYLINDRICAL

Figure 1.6.1 Classification by Projection Surface

1.7. Classification by Orientation of the Azimuthal Plane [24]

Azimuthal projections may be classified by reference to the point of contact of the plotting surface with the earth. Azimuthal projections may be classified as polar, equatorial, or oblique. When the plane is tangent to the earth at either pole, we have a polar projection. When the plane is tangent to the earth at any point on the equator, the projection is called equatorial. The oblique case occurs when the plane is tangent at any point on the earth except the poles and the equator. Figure 1.7.1 indicates these three alternatives. In each case, T is the point of tangency. OT is the line from the center of the earth to the point of tangency.

1.8 Classification by Orientation of a Cone or Cylinder [24]

Another classification set can be defined for cones and cylinders. These plotting surfaces may be considered to be regular, transverse, or oblique.

The regular projection occurs when the axis of the cone or cylinder coincides with the polar axis of the earth. The transverse case has the axis of the cone or cylinder perpendicular to, and intersecting the axis of the earth, in the plane of the equator. The transverse Mercator, and the transverse Polyconic are examples of this. If the axis of the cone or cylinder has any other position in space, besides being coincident with, or perpendicular to, the axis of the earth, then an oblique projection is generated. In all cases, the axis of the cone or cylinder passes through the center of the earth. Figure 1.8.1 demonstrates these three options for a cylindrical projection. In this figure the circle of tangency of the cylinder with the sphere is denoted by dotted lines.

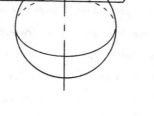

(a) POLAR

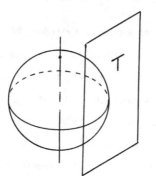

(b) EQUATORIAL

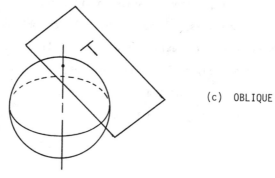

(c) OBLIQUE

Figure 1.7.1. Orientation of the Azimuthal Plane

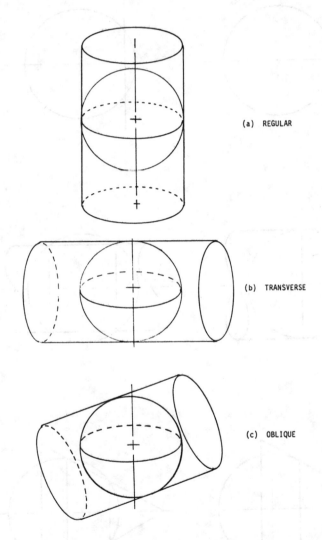

(a) REGULAR

(b) TRANSVERSE

(c) OBLIQUE

Figure 1.8.1   Orientation of a Cylinder

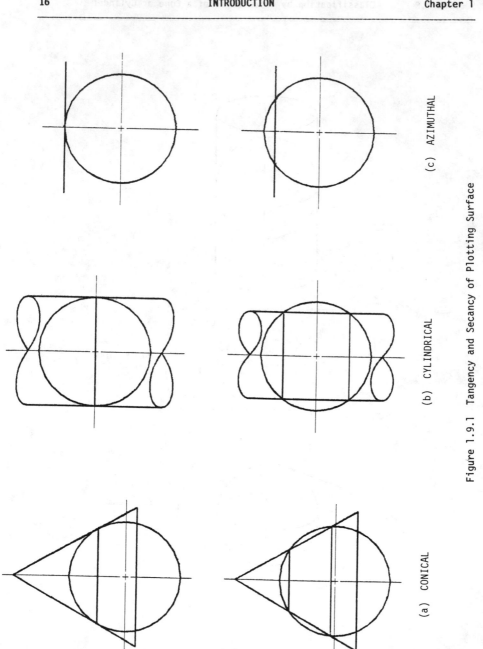

(a) CONICAL   (b) CYLINDRICAL   (c) AZIMUTHAL

Figure 1.9.1 Tangency and Secancy of Plotting Surface

1.9 Classification by Tangency or Secancy

Another classification set rests on whether the plotting surface is tangent or secant to the model of the earth. Figure 1.9.1 gives the comparison of tangent and secant plotting surfaces for the conical, cylindrical, and azimuthal plotting surfaces. Note that the tangent case results in one true length line, while the secant case results in two true length lines.

1.10 Projecton Technique [3], [11]

Three techniques of projection can be identified. This can serve as another scheme of classification. The methods are the graphical, the semi-graphical, and the mathematical.

In any graphical technique, some point 0 is chosen as a projection point, and the methods of projective geometry and descriptive geometry are used to transform a point P on the earth to a location P' on the plotting surface. An example of this is indicated in Figure 1.10.1 where the point P on the earth

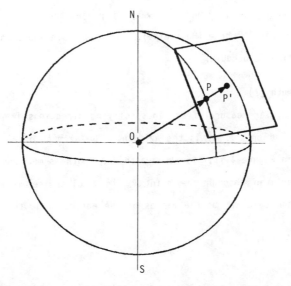

Figure 1.10.1 Graphical Projection Onto a Plane

is transformed to the oblique plane by the extension of line OP until it intersects the plane. In this example, O is arbitrarily chosen as the projection point. Since anything that can be done graphically can also be described mathematically, we do not encourage graphical constructions. However, those projections which are capable of a strict graphical approach are identified in Chapters 4, 5, and 6.

Those projections termed mathematical are those which can only be produced by a mathematical definition. No draftsman with compass and straight edge can plot them by means of the projection of a ray. The mathematical projections are often be characterized by an equation the form of:

$$\left\{ \begin{matrix} x \\ y \end{matrix} \right\} = [T] \left\{ \begin{matrix} \phi \\ \lambda \end{matrix} \right\} \qquad\qquad (1.10.1)$$

In this equation x and y are the Cartesian coordinates, and ϕ and λ are latitude and longitude, respectively. The generally non-constant matrix T is the transformation matrix, which is developed in Chapter 2, and applied.

In between these two groups are the semi-graphical projections. However, the various reasons, such as a varying projection point (Mercator), or a complex graphical scheme (Mollweide), the reasonable approach is to depend on a mathematical procedure.

1.11 Azimuth [7]

The angular measure of use in specifying directions in planes tangent to the earth and on the map is the azimuth. The azimuth of P' in relation to P is shown in Figure 1.11.1 for the earth. Azimuth is measured from the north, on the meridian through the point P, in a clockwise manner. Azimuth is measured the same way on the map as on the earth. Azimuth ranges from 0° to 360°.

1.12   Constant of a Cone

At this point it is convenient to introduce the concept of the constant of the cone. This is done for cones both tangent and secant to the earth. Consider first the tangent case. Let a be the radius of the earth. From Figure 1.12.1 the slant height of the cone tangent to the earth, ρ, is found

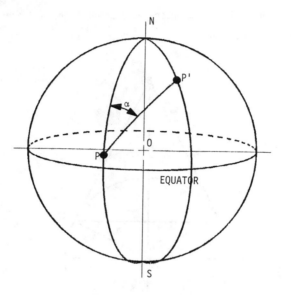

Figure 1.11.1   Azimuth of P' From P

to be

$$\rho = a \cot \phi \qquad\qquad (1.12.1)$$

where ϕ is the latitude. Also, from the figure, d, the circumference of the parallel circle AB, which defines the circle of tangency of the cone, is

$$d = 2\pi a \cos \phi. \qquad\qquad (1.12.2)$$

The constant of the cone, c, is defined from the relation between the developed cone and the earth. Let,

$$\theta = d/\rho. \qquad\qquad (1.12.3)$$

Substitute Eqs. (1.12.1) and (1.12.2) into Eq. (1.12.3)

$$\theta = \frac{2\pi a \cos \phi}{a \cot \phi}$$

$$\theta = 2\pi \sin \phi. \qquad\qquad\qquad (1.12.4)$$

The constant of the cone is $c = \sin \phi$. It is a multiplicative factor that relates longitudes on the earth to those on the cone. Equations (1.12.1) and (1.12.4) will be beneficial in Chapters 4, 5, and 6 in the investigation of the various conical projections.

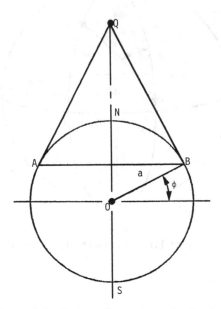

Figure 1.12.1   Cone Tangent to the Earth

Note in Eq. (1.12.4) that as ϕ varies from 0° to 90°, θ varies from 0° to 360°. When θ is 0°, then we have a cylinder. At θ equals 360°, we have a plane. As was mentioned above, it is useful, usually, to treat planes, cones and cylinders as separate entities, rather than lump them together in a single general approach to the problem.

The cone may also be secant to the earth. This is shown in Figure 1.12.2, where the circles of secancy are at the latitudes ϕ_1, and ϕ_2. From the similar triangles, the ratios of the slant heights of these respective latitudes are

$$\frac{\rho_1}{\rho_2} = \frac{a \cos \phi_1}{a \cos \phi_2}$$

$$= \frac{\cos \phi_1}{\cos \phi_2}. \tag{1.12.5}$$

Equation (1.12.5) is the basic relation used in Chapters 4, 5, and 6 to obtain the constant of the cone for conical projections with two standard parallels under specific requirements.

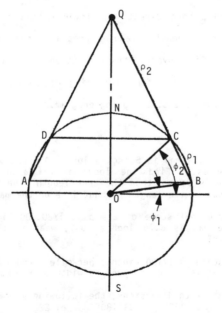

Figure 1.12.2 Cone Secant to the Earth

1.13 Plotting Equations [12], [13]

The subject of map projections certainly has intrinsic interest. Of more importance to surveyors, cartographers, and all other workers in the field of map projections, is the availability of plotting equations. The ultimate goal of this text is to derive cartesian plotting equations as a function of latitude and longitude for each direct transformal. This is, at the very least, a good beginning for further practical work. After the basic concepts are derived in Chapters 2 and 3, the actual equations for mapping are developed in Chapters 4, 5, and 6.

Of nearly equal importance to workers in the field of mapping are the inverse transformations, from cartesian coordinates to geographic. These too are developed in Chapters 4, 5, and 6 for selected map projections.

The transformations, both direct and inverse, can be programmed for inclusion in either a comprehensive mapping program, or as subroutines in a specialized program. The inverse transformation can be of particular importance in computer graphics for digitizing of an existing map. See Reference [20] for one comprehensive mapping program.

PROBLEMS

1.1 A true length line on a map is 5 inches long. A corresponding distorted line is 4.93 inches. Find the local scale factor. If a third corresponding line is longer than the true length line by a local scale factor of 1.05, what is the length of that line on the map?

1.2 Given that the principal scale of a map is 1/20,000 (inches per feet). If a line on the map is 6.75 inches long, what is the corresponding length on the earth?

1.3 If the principal scale is 1/100 (inches per mile), and 750 miles in width will be mapped, what is the corresponding width on the map?

1.4 Sketch, and indicate with dimensions, the following azimuths:
a) 0°   b) 45°   c) 135°   d) 180°   e) 225°   f) 315°

1.5 For a cone tangent to a sphere at latitude 45°, what is the constant of the cone?

1.6 For a cone secant to a sphere at latitudes 30° and 40°, what is the ratio of the slant heights?

2

Mapping Transformations

The process of map projection requires the transformation from the two independent coordinates of the earth to the two independent coordinates of the map. This chapter is devoted to the general theory of transformations. To this end, it is first necessary to develop some applicable formulas of differential geometry, and apply some aspects of spherical trigonometry.

The differential geometry of curves is discussed to introduce needed concepts, such as radius of curvature and torsion of a space curve. The differential geometry of surfaces discussion introduces the first and second fundamental forms, and parametric curves and the condition of orthogonality. The surfaces of interest in mapping are surfaces of revolution. The general surface is particularized to surfaces of revolution (Chapter 3). The process of transformation from non-developable to developable surfaces is considered. Representations of arc length, angles, and area, as well as the definition of the normal to the surface, are also discussed.

The basic transformation matrix introduced in Section 1.10 is derived in this chapter. The conditions of equal area and conformality are applied to this transformation. The convergence of the meridians is then considered. Finally, a rotation method for the production of equatorial, transverse, and oblique projections is discussed.

2.1 Differential Geometry of Curves [10]

Consider the space curve of Figure 2.1.1. Let ζ be an arbitrary parameter. Let the vector to any point P, on the curve, in the cartesian coordinate system, be

$$\mathbf{r} = x(\zeta)\hat{i} + y(\zeta)\hat{j} + z(\zeta)\hat{k}. \qquad (2.1.1)$$

Let $|\Delta\mathbf{r}| = \Delta s$.

The unit tangent vector at point P is

$$\lim_{\Delta s \to 0} \frac{\Delta \mathbf{r}}{\Delta s} = \frac{d\mathbf{r}}{ds} = \hat{t}.$$

(2.1.2)

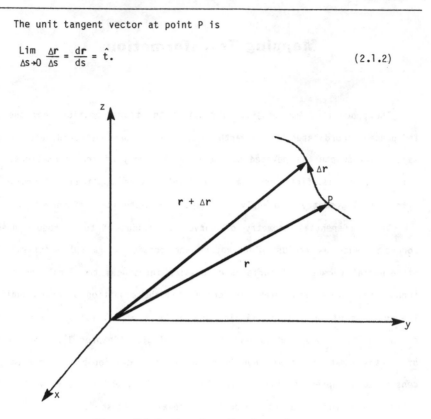

Figure 2.1.1 Geometry of a Space Curve

Applying the chain rule to Eq. (2.1.2), one finds

$$\hat{t} = \frac{d\mathbf{r}}{d\zeta} \frac{d\zeta}{ds}.$$

(2.1.3)

Taking the total differential of Eq. (2.1.1), we note

$$\hat{t} = \left(\frac{\partial x}{\partial \zeta} \hat{i} + \frac{\partial y}{\partial \zeta} \hat{j} + \frac{\partial z}{\partial \zeta} \hat{k} \right) \left(\frac{d\zeta}{ds} \right).$$

(2.1.4)

Upon taking the dot product of $\hat{t}$ with itself, we have

$$\hat{t} \cdot \hat{t} = 1 = \left[\left(\frac{\partial x}{\partial \zeta} \right)^2 + \left(\frac{\partial y}{\partial \zeta} \right)^2 + \left(\frac{\partial z}{\partial \zeta} \right)^2 \right] \left(\frac{d\zeta}{ds} \right)^2$$

$$\left(\frac{d\zeta}{ds} \right)^2 = \frac{1}{\left(\frac{\partial x}{\partial \zeta} \right)^2 + \left(\frac{\partial y}{\partial \zeta} \right)^2 + \left(\frac{\partial z}{\partial \zeta} \right)^2}$$

$$\frac{d\zeta}{ds} = \frac{1}{\sqrt{(\frac{\partial x}{\partial \zeta})^2 + (\frac{\partial y}{\partial \zeta})^2 + (\frac{\partial z}{\partial \zeta})^2}}$$

$$= \frac{1}{|\frac{\partial \mathbf{r}}{\partial \zeta}|}. \qquad (2.1.5)$$

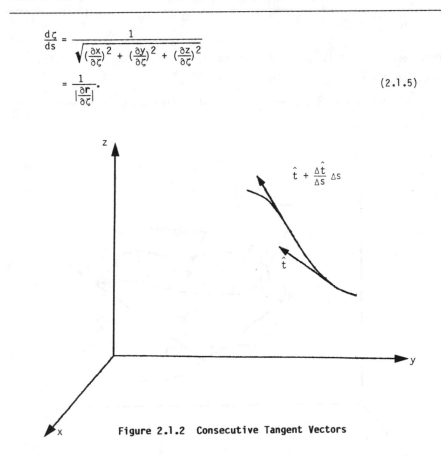

Figure 2.1.2   Consecutive Tangent Vectors

Next, we look at two consecutive tangent vectors, as shown in Figure 2.1.2.

$$\lim_{\Delta s \to 0} \frac{\Delta \hat{t}}{\Delta s} = \frac{d\hat{t}}{ds}$$

Let

$$\frac{d\hat{t}}{ds} = -k\hat{n} \qquad (2.1.6)$$

where k is defined as the curvature, and $\hat{n}$ is the principal unit normal.

Upon dotting $\hat{t}$ with itself, and differentiating, we find

$$\hat{t} \cdot \hat{t} = 1$$

$$2\hat{t} \cdot \frac{d\hat{t}}{ds} = 0.$$

This means that $\hat{t}$ is perpendicular to $d\hat{t}/ds$, and, from Eq. (2.1.6), $\hat{n}$ is perpendicular to $\hat{t}$.

In order to obtain a right-handed triad, define the binormal vector

$$\hat{b} = \hat{t} \times \hat{n}. \qquad\qquad\qquad\qquad\qquad (2.1.7)$$

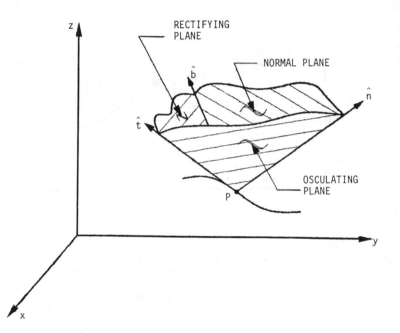

Figure 2.1.3 Planes on the Space Curve at Point P

Thus, we have the fundamental set of unit vectors $\hat{t}$, $\hat{n}$, and $\hat{b}$ associated with a continuous space curve at point P.

It is useful at this time to define three types of planes intersecting the curve. These are the osculating, normal, and rectifying planes. The osculating plane contains $\hat{t}$ and $\hat{n}$. The normal plane contains $\hat{n}$ and $\hat{b}$. The rectifying plane is defined by $\hat{t}$ and $\hat{b}$. These planes are displayed in Figure 2.1.3.

It is useful now to obtain the derivatives of the unit vectors as a function of distance along the curve. From the definition of the unit vectors, we have the relations

$$\left.\begin{array}{l} \hat{b} = \hat{t} \times \hat{n} \\ \hat{t} = \hat{n} \times \hat{b} \\ \hat{n} = \hat{b} \times \hat{t} \end{array}\right\} . \qquad\qquad (2.1.8)$$

From the first of Eqs. (2.1.8)

$$\frac{d\hat{b}}{ds} = \frac{d}{ds} (\hat{t} \times \hat{n})$$

$$= \frac{d\hat{t}}{ds} \times \hat{n} + \hat{t} \times \frac{d\hat{n}}{ds}. \qquad\qquad (2.1.9)$$

Substitute Eq. (2.1.6) into Eq. (2.1.9), to obtain

$$\frac{d\hat{b}}{ds} = -k\hat{n} \times \hat{n} + \hat{t} \times \frac{d\hat{n}}{ds}$$

$$= \hat{t} \times \frac{d\hat{n}}{ds}. \qquad\qquad (2.1.10)$$

Dot $\hat{n}$ with itself, and differentiate, so that

$$\hat{n} \cdot \hat{n} = 1$$

$$2\hat{n} \cdot \frac{d\hat{n}}{ds} = 0.$$

Thus, $d\hat{n}/ds$ is perpendicular to $\hat{n}$, and must lie in the rectifying plane, and has the components

$$\frac{d\hat{n}}{ds} = \phi\hat{t} + \tau\hat{b}. \qquad\qquad (2.1.11)$$

Substituting Eq. (2.1.11) into Eq. (2.1.10), we find

$$\frac{d\hat{b}}{ds} = \hat{t} \times (\phi t + \tau\hat{b})$$

$$= \tau\hat{t} \times \hat{b}$$

$$= -\tau\hat{n}. \qquad\qquad (2.1.12)$$

The constant τ is called the torsion. It is, essentially, a measure of the twist of the curve out of the osculating plane.

From the last of (2.1.8)

$$\frac{d\hat{n}}{ds} = \frac{d}{ds}(\hat{b} \times \hat{t})$$

$$= \frac{d\hat{b}}{ds} \times \hat{t} + \hat{b} \times \frac{d\hat{t}}{ds}. \qquad (2.1.13)$$

Substitute Eqs. (2.1.6) and (2.1.12) into Eq. (2.1.13); this gives

$$\frac{d\hat{n}}{ds} = -\tau\hat{n} \times \hat{t} + \hat{b} \times (-k\hat{n})$$

$$= \tau\hat{b} + k\hat{t}. \qquad (2.1.14)$$

Equations (2.1.6), (2.1.13), and (2.1.14) can be arranged in matrix form.

$$\begin{Bmatrix} d\hat{t}/ds \\ d\hat{n}/ds \\ d\hat{b}/ds \end{Bmatrix} = \begin{bmatrix} 0 & -k & 0 \\ k & 0 & \tau \\ 0 & -\tau & 0 \end{bmatrix} \begin{Bmatrix} \hat{t} \\ \hat{n} \\ \hat{b} \end{Bmatrix}. \qquad (2.1.15)$$

These are the Frenet-Serret formulas.

The next step is to obtain the mathematical relations for the curvature and the torsion.

The curvature, in general parametric form, is obtained from Eqs. (2.1.2) and (2.1.6).

$$-k\hat{n} = \frac{d\hat{t}}{ds}$$

$$= \frac{d}{ds}\left(\frac{d\mathbf{r}}{ds}\right)$$

$$= \frac{d^2\mathbf{r}}{ds^2} \qquad (2.1.16)$$

Taking the cross product of $\hat{t}$ with Eq. (2.1.16), we have

$$\hat{t} \times (-k\hat{n}) = \hat{t} \times \left(\frac{d^2\mathbf{r}}{ds^2}\right). \qquad (2.1.17)$$

Using the first of Eqs. (2.1.8), and (2.1.2) in Eq. (2.1.17),

$$-k\hat{b} = \frac{d\mathbf{r}}{ds} \times \frac{d^2\mathbf{r}}{ds^2}$$

$$k = |\frac{d\mathbf{r}}{ds} \times \frac{d^2\mathbf{r}}{ds^2}|. \qquad (2.1.18)$$

For the general parametrization, we have the **r** derivatives

$$\frac{d\mathbf{r}}{ds} = \frac{d\mathbf{r}}{d\zeta} \frac{d\zeta}{ds} \qquad (2.1.19)$$

$$\frac{d^2\mathbf{r}}{ds^2} = \frac{d^2\mathbf{r}}{d\zeta^2} (\frac{d\zeta}{ds})^2 + \frac{d\mathbf{r}}{d\zeta} (\frac{d^2\zeta}{ds^2}). \qquad (2.1.20)$$

Note, substituting Eqs. (2.1.19) and (2.1.20) into Eq. (2.1.18), that

$$k = |\frac{d\mathbf{r}}{d\zeta} \frac{d\zeta}{ds} \times [\frac{d^2\mathbf{r}}{d\zeta^2} (\frac{d\zeta}{ds})^2 + \frac{d\mathbf{r}}{d\zeta} (\frac{d^2\zeta}{ds^2})]|$$

$$= |\frac{d\mathbf{r}}{d\zeta} \times \frac{d^2\mathbf{r}}{d\zeta^2}| (\frac{d\zeta}{ds})^3. \qquad (2.1.21)$$

Upon substituting Eq. (2.1.5) into Eq. (2.1.21), we arrive at

$$k = \frac{|\frac{d\mathbf{r}}{d\zeta} \times \frac{d^2\mathbf{r}}{d\zeta^2}|}{|\frac{d\mathbf{r}}{d\zeta}|^3}. \qquad (2.1.22)$$

A similar procedure can be followed to obtain the torsion. Upon taking the dot product of Eq. (2.1.12) with $\hat{n}$.

$$\tau = -\frac{d\hat{b}}{ds} \cdot \hat{n}. \qquad (2.1.23)$$

Substitute the first of Eqs. (2.1.8) into Eq. (2.1.23), this gives

$$\tau = -\frac{d}{ds} (\hat{t} \times \hat{n}) \cdot \hat{n}$$

$$= -(\frac{d\hat{t}}{ds} \times \hat{n} + \hat{t} \times \frac{d\hat{n}}{ds}) \cdot \hat{n}$$

$$= (\hat{t} \times \hat{n}) \cdot \frac{d\hat{n}}{ds}. \qquad (2.1.24)$$

From Eq. (2.1.16)

$$\hat{n} = - \frac{\dfrac{d^2\mathbf{r}}{ds^2}}{k} \tag{2.1.25}$$

$$\frac{d\hat{n}}{ds} = - \frac{\dfrac{d^3\mathbf{r}}{ds^3}}{k} + \frac{\dfrac{dk}{ds}}{k^2} \frac{d^2\mathbf{r}}{ds^2}. \tag{2.1.26}$$

Substitute Eqs. (2.1.2), (2.1.25), and (2.1.26) into Eq. (2.1.24), to obtain

$$\tau = \quad - \frac{d\mathbf{r}}{ds} \times \frac{\dfrac{d^2\mathbf{r}}{ds^2}}{k} \quad \cdot \quad \frac{\dfrac{d^3\mathbf{r}}{ds^3}}{k} + \frac{\dfrac{dk}{ds}}{k^2} \frac{d^2\mathbf{r}}{ds^2}$$

$$= \frac{1}{k^2} \left(\frac{d\mathbf{r}}{ds} \times \frac{d^2\mathbf{r}}{ds^2}\right) \cdot \frac{d^2\mathbf{r}}{ds^2}. \tag{2.1.27}$$

For the general parameterization, differentiate Eq. (2.1.20)

$$\frac{d^3\mathbf{r}}{ds^3} = \frac{d^3\mathbf{r}}{d\zeta^3} \left(\frac{d\zeta}{ds}\right)^3 + 3 \frac{d^2\mathbf{r}}{d\zeta^2} \left(\frac{d\zeta}{ds}\right)\left(\frac{d^2\zeta}{ds^2}\right)$$

$$+ \frac{d\mathbf{r}}{d\zeta} \frac{d^3\zeta}{ds^3}. \tag{2.1.28}$$

Substitute Eqs. (2.1.19), (2.1.20), and (2.1.28) into Eq. (2.1.27) to obtain

$$\tau = \frac{[(\frac{d\mathbf{r}}{d\zeta} \times \frac{d^2\mathbf{r}}{d\zeta^2}) \cdot (\frac{d^3\mathbf{r}}{d\zeta^3})](\frac{d\zeta}{ds})^6}{k^2}. \tag{2.1.29}$$

Substituting Eq. (2.1.5) into Eq. (2.1.29), then Eq. (2.1.29) simplifies to

$$\tau = \frac{[(\frac{d\mathbf{r}}{d\zeta} \times \frac{d^2\mathbf{r}}{d\zeta^2}) \cdot (\frac{d^3\mathbf{r}}{d\zeta^3})]}{k^2 |\frac{d\mathbf{r}}{d\zeta}|^6}. \tag{2.1.30}$$

The torsion is important for such curves as the geodesic (Chapter 3). For plane curves, such as the meridian curve, and the equator, $\tau = 0$.

As an example, consider a plane curve [16], and let $\zeta = x$, and $y = y(x)$. The radius vector is

$$\mathbf{r} = x\hat{i} + y(x)\hat{j}. \tag{2.1.31}$$

Obtain the curvature, by differentiating Eq. (2.1.31).

$$\frac{d\mathbf{r}}{dx} = \hat{i} + \frac{dy}{dx} \hat{j} \qquad (2.1.32)$$

$$\frac{d^2\mathbf{r}}{dx^2} = \frac{d^2y}{dx^2} \hat{j}. \qquad (2.1.33)$$

Substituting Eq. (2.1.32) and (2.1.33) into Eq. (2.1.22), we have

$$k = \frac{|(\hat{i} + \frac{dy}{dx} \hat{j}) \times (\frac{d^2y}{dx^2} \hat{j})|}{[(\hat{i} + \frac{dy}{dx} \hat{j}) \cdot (\hat{i} + \frac{dy}{dx} \hat{j})]^{3/2}}$$

$$= \frac{|\frac{d^2y}{dx^2}| \hat{k}}{[1 + (\frac{dy}{dx})^2]^{3/2}}. \qquad (2.1.34)$$

The radius of curvature is the reciprocal of the curvature. Thus, from Eq. (2.1.34), and taking the magnitude,

$$\rho = 1/k$$

$$= \frac{[1 + (\frac{dy}{dx})^2]^{3/2}}{\frac{d^2y}{dx^2}} . \qquad (2.1.35)$$

Continuing from Eq. (2.1.33), we note that

$$\frac{d^3\mathbf{r}}{dx^3} = \frac{d^3y}{dx^3} \hat{j}. \qquad (2.1.36)$$

Substituting Eqs. (2.1.32), (2.1.33), and (2.1.36) into Eq. (2.1.30).

$$\tau = \frac{\begin{vmatrix} 1 & \frac{dy}{dx} & 0 \\ 0 & \frac{d^2y}{dx^2} & 0 \\ 0 & \frac{d^3y}{dx^3} & 0 \end{vmatrix}}{k^2 |\hat{i} + \frac{dy}{dx} \hat{j}|^6} = 0.$$

2.2 Differential Geometry of Surfaces [10]

The parametric representation of a surface requires two parameters. In general, for the parametric representation of a surface by two arbitrary parameters, α_1 and α_2, the vector to a point on the surface is

$$\mathbf{r} = \mathbf{r}(\alpha_1, \alpha_2). \qquad (2.2.1)$$

If either of the two parameters is held constant, and the other one is varied, a space curve results. This space curve is the parametric curve. Figure 2.2.1 gives the parametric representation of space curves on a surface. The α_1-curve is the parametric curve along which α_2 is constant, and the α_2-curve is the parametric curve along which α_1 is constant.

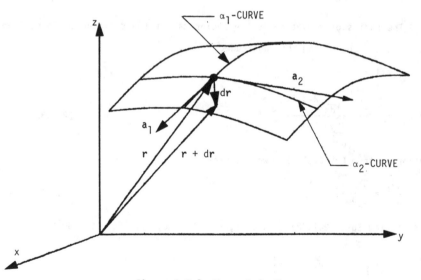

Figure 2.2.1 Parametric Curves

The next step is to obtain the tangents to the parametric curves at point P. The tangent vector to the α_1-curve is

$$\mathbf{a}_1 = \frac{\partial \mathbf{r}}{\partial \alpha_1}. \qquad (2.2.2)$$

The tangent to the α_2-curve is

$$\mathbf{a}_2 = \frac{\partial \mathbf{r}}{\partial \alpha_2}. \tag{2.2.3}$$

The plane spanned by the vectors $\mathbf{a}_1$ and $\mathbf{a}_2$ is the tangent plane to the surface at point P.

The total differential of Eq. (2.2.1) is

$$d\mathbf{r} = \frac{\partial \mathbf{r}}{\partial \alpha_1} d\alpha_1 + \frac{\partial \mathbf{r}}{\partial \alpha_2} d\alpha_2. \tag{2.2.4}$$

Substituting Eqs. (2.2.2) and (2.2.3) into Eq. (2.2.4), we have

$$d\mathbf{r} = \mathbf{a}_1 \, d\alpha_1 + \mathbf{a}_2 \, d\alpha_2. \tag{2.2.5}$$

Armed with Eq. (2.2.5), we are now ready to introduce the first fundamental form.

2.3 First Fundamental Form [10]

The first fundamental form of a surface is now to be derived. The first fundamental form is useful in dealing with arc length, area, angular measure on a surface, and the normal to the surface. Take the dot product of Eq. (2.2.5) with itself to obtain a scalar.

$$
\begin{aligned}
(ds)^2 &= d\mathbf{r} \cdot d\mathbf{r} \\
&= (\mathbf{a}_1 \, d\alpha_1 + \mathbf{a}_2 \, d\alpha_2) \cdot (\mathbf{a}_1 \, d\alpha_1 + \mathbf{a}_2 \, d\alpha_2) \\
&= \mathbf{a}_1 \cdot \mathbf{a}_1 (d\alpha_1)^2 + 2(\mathbf{a}_1 \cdot \mathbf{a}_2) d\alpha_1 \, d\alpha_2 \\
&\quad + \mathbf{a}_2 \cdot \mathbf{a}_2 (d\alpha_2)^2.
\end{aligned} \tag{2.3.1}
$$

Define new variables.

$$\left. \begin{aligned} E &= \mathbf{a}_1 \cdot \mathbf{a}_1 \\ F &= \mathbf{a}_1 \cdot \mathbf{a}_2 \\ G &= \mathbf{a}_2 \cdot \mathbf{a}_2 \end{aligned} \right\}. \tag{2.3.2}$$

Substituting Eq. (2.3.2) into Eq. (2.3.1), we obtain the important differential form

$$(ds)^2 = E(d\alpha_1)^2 + 2F d\alpha_1 d\alpha_2 + G(d\alpha_2)^2. \tag{2.3.3}$$

Equation (2.3.3) is the first fundamental form of a surface, and this is very useful through the whole process of map projection. The first fundamental form is now applied to linear measure on any surface.

Arc length can be found immediately from the integration of Eq. (2.3.3). The distance between two arbitrary points P_1 and P_2 on the surface is given by

$$s = \int_{P_1}^{P_2} \sqrt{E(d\alpha_1)^2 + 2Fd\alpha_1 d\alpha_2 + G(d\alpha_2)^2}$$
$$= \int_{P_1}^{P_2} \left\{ \sqrt{E + 2F\left(\frac{d\alpha_2}{d\alpha_1}\right) + G\left(\frac{d\alpha_2}{d\alpha_1}\right)^2} \right\} d\alpha_1. \tag{2.3.4}$$

Equation (2.3.4) is useful as soon as $d\alpha_2/d\alpha_1$ is defined, and is used in Chapter 3 for distance along the spheroid.

Angles between two unit tangents $\mathbf{a}_1$ and $\mathbf{a}_2$ on the surface can be found by taking the dot product of Eqs. (2.2.2) and (2.2.3) and applying Eq. (2.3.2).

$$\cos\theta = \frac{\mathbf{a}_1}{|\mathbf{a}_1|} \cdot \frac{\mathbf{a}_2}{|\mathbf{a}_2|} \tag{2.3.5}$$

$$= \frac{F}{\sqrt{EG}} \tag{2.3.6}$$

$$\sin\theta = \sqrt{1 - \cos^2\theta}$$

$$= \sqrt{1 - \frac{F^2}{EG}}$$

$$= \sqrt{\frac{EG - F^2}{EG}}. \tag{2.3.7}$$

Define
$$H = EG - F^2 \tag{2.3.8}$$
and substitute Eq. (2.3.8) into Eq. (2.3.7), to obtain

$$\sin\theta = \sqrt{\frac{H}{EG}}.$$

The normal to the surface at point P is

$$\hat{n} = \frac{a_1 \times a_2}{|a_1 \times a_2|}$$

$$= \frac{a_1 \times a_2}{|a_1||a_2|\sin\theta}. \tag{2.3.10}$$

Substituting Eqs. (2.3.2) and (2.3.8) into Eq. (2.3.9), we have

$$\hat{n} = \frac{a_1 \times a_2}{\sqrt{E}\,\sqrt{G}\,\sqrt{\frac{H}{EG}}}$$

$$= \frac{a_1 \times a_2}{\sqrt{H}}. \tag{2.3.11}$$

Incremental area can be obtained by a consideration of incremental distance along the parametric curves. Along the α_1-curve, and the α_1-curve, respectively,

$$ds_1 = \sqrt{E}\,d\alpha_1$$
$$ds_2 = \sqrt{G}\,d\alpha_2 \tag{2.3.12}$$

The differential area is then

$$dA = ds_1 ds_2 \sin\theta$$

$$= \sqrt{EG}\,d\alpha_1 d\alpha_2 \sin\theta. \tag{2.3.13}$$

Substituting Eq. (2.3.8) into Eq. (2.3.13), we have

$$dA = \sqrt{EG}\,\sqrt{\frac{H}{EG}}\,d\alpha_1 d\alpha_2$$

$$= \sqrt{H}\,d\alpha_1 d\alpha_2. \tag{2.3.14}$$

Thus, the first fundamental form has given a means to derive the arc length, the unit normal to the surface at every point, and incremental area. In conjunction with the second fundamental form of the next section, it will be useful in determining the radii of curvature of the surface.

As is subsequently shown in Chapter 3, the first fundamental form for the sphere is

$$(ds)^2 = a^2(d\phi)^2 + a^2\cos^2\phi(d\lambda)^2 \tag{2.3.15}$$

and for the spheroid, it is

$$(ds)^2 = R_m^2 (d\phi)^2 + R_p^2 \cos^2 \phi (d\lambda)^2. \tag{2.3.16}$$

When the chosen parameters are such as to ensure that the parametric curves are orthogonal to each other, a simplification of the first fundamental form occurs. When orthogonality is present, from Eq. (2.3.2), a_1 and a_2 are perpendicular, and F = 0. The first fundamental form is then

$$(ds)^2 = E(d\alpha_1)^2 + G(d\alpha_2)^2. \tag{2.3.17}$$

2.4 The Second Fundamental Form [10]

The second fundamental form provides a way to evaluate principal directions and curvatures of the surface. We deal below with normal sections through the surface, and derive formulas for the curvature of a normal section. A normal section implies that the normal to the parametric curve and the surface coincide.

We begin with a development to obtain the formula for curvature, starting with the normal at an arbitrary point. For the parametric curve, take the dot product of $\hat{n}$ with Eq. (2.1.6)

$$\hat{n} \cdot \frac{d\hat{t}}{ds} = -k\hat{n} \cdot \hat{n}$$

$$= -k. \tag{2.4.1}$$

Substitute the derivative of Eq. (2.1.2) into Eq. (2.4.1), to obtain

$$k = - \frac{d^2 r}{ds^2} \cdot \hat{n}. \tag{2.4.2}$$

Since $\hat{t}$ and $\hat{n}$ are orthogonal,

$$\hat{t} \cdot \hat{n} = 0. \tag{2.4.3}$$

Then, substituting Eq. (2.1.2) into Eq. (2.4.3), we see

$$\frac{dr}{ds} \cdot \hat{n} = 0. \tag{2.4.4}$$

Upon taking the derivative of Eq. (2.4.4), we obtain

$$\frac{d}{ds}\left(\frac{d\mathbf{r}}{ds} \cdot \hat{n}\right) = 0$$

$$= \frac{d^2\mathbf{r}}{ds^2} \cdot \hat{n} + \frac{d\mathbf{r}}{ds} \cdot \frac{d\hat{n}}{ds}$$

$$\frac{d\mathbf{r}}{ds} \cdot \frac{d\hat{n}}{ds} = -\frac{d^2\mathbf{r}}{ds^2} \cdot \hat{n}. \qquad (2.4.5)$$

Substitute Eq. (2.4.2) into Eq. (2.4.5), this gives

$$\frac{d\mathbf{r}}{ds} \cdot \frac{d\hat{n}}{ds} = k. \qquad (2.4.6)$$

From the total differential of $\mathbf{r}$ and $\hat{n}$, we have

$$d\mathbf{r} = \frac{\partial\mathbf{r}}{\partial\alpha_1}d\alpha_1 + \frac{\partial\mathbf{r}}{\partial\alpha_2}d\alpha_2 \qquad (2.4.7)$$

$$d\hat{n} = \frac{\partial\hat{n}}{\partial\alpha_1}d\alpha_1 + \frac{\partial\hat{n}}{\partial\alpha_2}d\alpha_2. \qquad (2.4.8)$$

Substituting Eqs. (2.4.7) and (2.4.8) into Eq. (2.4.6) gives

$$k = \frac{\left(\frac{\partial\mathbf{r}}{\partial\alpha_1}d\alpha_1 + \frac{\partial\mathbf{r}}{\partial\alpha_2}d\alpha_2\right) \cdot \left(\frac{\partial\hat{n}}{\partial\alpha_1}d\alpha_1 + \frac{\partial\hat{n}}{\partial\alpha_2}d\alpha_2\right)}{(ds)^2} \qquad (2.4.9)$$

Finally, substituting Eqs. (2.2.2), (2.2.3), and (2.3.3) into Eq. (2.4.9), we obtain

$$k = \frac{\mathbf{a}_1 \cdot \frac{\partial\hat{n}}{\partial\alpha_1}(d\alpha_1)^2 + \mathbf{a}_2 \cdot \frac{\partial\hat{n}}{\partial\alpha_1}(d\alpha_2)^2 + \left(\mathbf{a}_1 \cdot \frac{\partial\hat{n}}{\partial\alpha_2} + \mathbf{a}_2 \cdot \frac{\partial\hat{n}}{\partial\alpha_1}\right)d\alpha_1 d\alpha_2}{E(d\alpha_1)^2 + 2Fd\alpha_1 d\alpha_2 + G(d\alpha_2)^2}$$

$$(2.4.10)$$

The second fundamental form is defined as

$$\mathbf{a}_1 \cdot \frac{\partial\hat{n}}{\partial\alpha_1}(d\alpha_1)^2 + \mathbf{a}_2 \cdot \frac{\partial\hat{n}}{\partial\alpha_2}(d\alpha_2)^2$$

$$+ \left(\mathbf{a}_1 \cdot \frac{\partial\hat{n}}{\partial\alpha_2} + \mathbf{a}_2 \cdot \frac{\partial\hat{n}}{\partial\alpha_1}\right)d\alpha_1 d\alpha_2. \qquad (2.4.11)$$

Thus Eq. (2.4.10) is the ratio

$$k = \frac{\text{Second fundamental form}}{\text{First fundamental form}}.$$

It remains to define the coefficients of the differntials in the second fundamental form. From the definitions of the tangent and normal vectors

$$\hat{n} \cdot a_i = \hat{n} \cdot \frac{\partial r}{\partial \alpha_i}$$

$$= 0. \tag{2.4.12}$$

Taking the derivative of Eq. (2.4.12), we have

$$\frac{\partial}{\partial \alpha_j} (\hat{n} \cdot \frac{\partial r}{\partial \alpha_i}) = 0$$

$$\frac{\partial \hat{n}}{\partial \alpha_j} \cdot \frac{\partial r}{\partial \alpha_i} + \hat{n} \cdot \frac{\partial^2 r}{\partial \alpha_i \partial \alpha_j} = 0. \tag{2.4.13}$$

By definition, the second fundamental quantities are

$$b_{ij} = \frac{\partial \hat{n}}{\partial \alpha_j} \cdot \frac{\partial r}{\partial \alpha_i}. \tag{2.4.14}$$

Substitute Eq. (2.4.14) into Eq. (2.4.13), to obtain

$$b_{ij} = - \hat{n} \cdot \frac{\partial^2 r}{\partial \alpha_i \partial \alpha_j}. \tag{2.4.15}$$

Substituting Eq. (2.3.9) into Eq. (2.4.15), we have

$$b_{ij} = - \frac{a_1 \times a_2}{\sqrt{H}} \cdot \frac{\partial^2 r}{\partial \alpha_i \partial \alpha_j}$$

$$= - \frac{1}{\sqrt{H}} (\frac{\partial^2 r}{\partial \alpha_i \partial \alpha_j} \times a_1) \cdot a_2$$

$$= - \frac{1}{\sqrt{H}} \begin{vmatrix} \frac{\partial^2 x}{\partial \alpha_i \partial \alpha_j} & \frac{\partial^2 y}{\partial \alpha_i \partial \alpha_j} & \frac{\partial^2 z}{\partial \alpha_i \partial \alpha_j} \\ \frac{\partial x}{\partial \alpha_1} & \frac{\partial y}{\partial \alpha_1} & \frac{\partial z}{\partial \alpha_1} \\ \frac{\partial x}{\partial \alpha_2} & \frac{\partial y}{\partial \alpha_2} & \frac{\partial z}{\partial \alpha_2} \end{vmatrix} \tag{2.4.16}$$

Defining

$$\left. \begin{aligned} L &= b_{11} \\ M &= b_{12} = b_{21} \\ N &= b_{22} \end{aligned} \right\} . \tag{2.4.17}$$

and substituting Eq. (2.4.17) into Eq. (2.4.10), we arrive at the result

$$k = \frac{L(d\alpha_1)^2 + 2Md\alpha_1 d\alpha_2 + N(d\alpha_2)^2}{E(d\alpha_1)^2 + 2Fd\alpha_1 d\alpha_2 + G(d\alpha_2)^2}. \tag{2.4.18}$$

We now have the curvature in terms of the first and second fundamental forms. The next step is to maximize Eq. (2.4.18) to obtain the principal directions. Let $\alpha_2 = \alpha_2(\alpha_1)$ and $\lambda = d\alpha_2/d\alpha_1$ where λ is an unspecified parametric directon. From Eq. (2.4.18)

$$k = \frac{L + 2M(\frac{d\alpha_2}{d\alpha_1}) + N(\frac{d\alpha_2}{d\alpha_1})^2}{E + 2F(\frac{d\alpha_2}{d\alpha_1}) + G(\frac{d\alpha_2}{d\alpha_1})^2}$$

$$= \frac{L + 2M\lambda + N\lambda^2}{E + 2F\lambda + G\lambda^2}. \tag{2.4.19}$$

To find the directions for which k is an extremum, take the derivative of Eq. (2.4.19) with respect to λ, and set this equal to zero.

$$\frac{dk}{d\lambda} = \frac{(2M + 2N\lambda)}{(E + 2F\lambda + G\lambda^2)} - \frac{(L + 2M\lambda + N\lambda^2)(2F + 2G\lambda)}{(E + 2F\lambda + G\lambda^2)^2}$$

$$= 0. \tag{2.4.20}$$

Substitute Eq. (2.4.19) into Eq. (2.4.20), to obtain

$$\frac{dk}{d\lambda} = \frac{(2M + 2N\lambda)}{(E + 2F\lambda + G\lambda^2)} - \frac{k(2F + 2G\lambda)}{(E + 2F\lambda + G\lambda^2)}$$

$$= 0.$$

Since the denominator will never be zero, we have the necessary condition that

$$2M + 2N\lambda - k(2F + 2G\lambda) = 0$$

$$k = \frac{M + N\lambda}{F + G\lambda}. \tag{2.4.21}$$

Write Eq. (2.4.19) as

$$k = \frac{(L + M\lambda) + \lambda(M + N\lambda)}{(E + F\lambda) + \lambda(F + G\lambda)}$$

$$k[(E + F\lambda) + \lambda(F + G\lambda)] = (L + M\lambda) + \lambda(M + N\lambda). \qquad (2.4.22)$$

and substitute Eq. (2.4.21) into Eq. (2.4.22); this gives

$$k[(E + F\lambda) + (F + G\lambda)\lambda] = (L + M\lambda) + \lambda(F + G\lambda)k$$

$$k(E + F\lambda) = L + M\lambda$$

$$k = \frac{L + M\lambda}{E + F\lambda}. \qquad (2.4.23)$$

Upon crossmultiplying Eqs. (2.4.21) and (2.4.23) to form a quadratic in λ (which will yield the principal directions), we obtain

$$(L + M\lambda)(F + G\lambda) = (M + N\lambda)(E + F\lambda)$$

$$LF + FM\lambda + GL\lambda + MG\lambda^2 = ME + MF\lambda + NE\lambda + NF\lambda^2$$

$$\lambda^2(MG - NF) + (LG - NE)\lambda + (LF - ME) = 0. \qquad (2.4.24)$$

The solutions of Eq. (2.4.24) are

$$\begin{Bmatrix} \lambda_1 \\ \lambda_2 \end{Bmatrix} = \frac{-(LG - NE) \pm \sqrt{(LG - NE)^2 - 4(MG - MF)(LF - ME)}}{2(MG - MF)}.$$
$$(2.4.25)$$

Apply the theory of equations for a quadratic to Eq. (2.4.24), to obtain

$$\lambda_1 + \lambda_2 = -\frac{(LG - NE)}{(MG - NF)} \qquad (2.4.26)$$

$$\lambda_1 \lambda_2 = \frac{(LF - ME)}{(MG - NF)}. \qquad (2.4.27)$$

The two principal directions will now be shown to be orthogonal. Let

$$\lambda_1 = \left(\frac{d\alpha_2}{d\alpha_1}\right) \qquad (2.4.28)$$

$$\lambda_2 = \left(\frac{\delta\alpha_2}{\delta\alpha_1}\right). \qquad (2.4.29)$$

Let θ be the angle between these two directions, and let d**r** and δ**r** be infinitesimal vectors along λ_1 and λ_2. The cosine of the angle between the vectors is

$$\cos \theta = \frac{d\mathbf{r}}{|d\mathbf{r}|} \cdot \frac{\delta\mathbf{r}}{|\delta\mathbf{r}|}$$

$$= \frac{d\mathbf{r}}{ds} \cdot \frac{\delta\mathbf{r}}{\delta s}. \tag{2.4.30}$$

The total differentials can be developed as follows.

$$d\mathbf{r} = \frac{\partial\mathbf{r}}{\partial\alpha_1} d\alpha_1 + \frac{\partial\mathbf{r}}{\partial\alpha_2} d\alpha_2 \tag{2.4.31}$$

$$\delta\mathbf{r} = \frac{\partial\mathbf{r}}{\partial\alpha_1} \delta\alpha_1 + \frac{\partial\mathbf{r}}{\partial\alpha_2} \delta\alpha_2. \tag{2.4.32}$$

Substitute Eqs. (2.4.31) and (2.4.32) into Eq. (2.4.30), and carry out the dot product, to obtain

$$\cos \theta = [(\frac{\partial\mathbf{r}}{\partial\alpha_1} \cdot \frac{\partial\mathbf{r}}{\partial\alpha_1}) d\alpha_1 \delta\alpha_1 + (\frac{\partial\mathbf{r}}{\partial\alpha_1} \cdot \frac{\partial\mathbf{r}}{\partial\alpha_2}) d\alpha_1 \delta\alpha_2$$

$$= (\frac{\partial\mathbf{r}}{\partial\alpha_2} \cdot \frac{\partial\mathbf{r}}{\partial\alpha_1}) d\alpha_2 \delta\alpha_1 + (\frac{\partial\mathbf{r}}{\partial\alpha_2} \cdot \frac{\partial\mathbf{r}}{\partial\alpha_2}) d\alpha_2 \delta\alpha_2] \times \frac{1}{ds\delta s} \tag{2.4.33}$$

Substitute Eq. (2.3.2) into Eq. (2.4.33), this yields

$$\cos \theta = \frac{1}{ds\delta s} [Ed\alpha_1 \delta\alpha_1 + F(d\alpha_1 \delta\alpha_2 + d\alpha_2 \delta\alpha_1) + Gd\alpha_2 \delta\alpha_2]$$

$$\frac{\cos \theta}{d\alpha_1 \delta\alpha_1} = \frac{1}{ds\delta s} [E + F(\frac{\delta\alpha_2}{\delta\alpha_1} + \frac{d\alpha_2}{d\alpha_1}) + G(\frac{\delta\alpha_2}{\delta\alpha_1})(\frac{d\alpha_2}{d\alpha_1})]. \tag{2.4.34}$$

Substitution of Eqs. (2.4.28) and (2.4.29) into Eq. (2.4.34) gives

$$\frac{\cos \theta}{\delta\alpha_1 d\alpha_1} = \frac{1}{ds\delta s} [E + F(\lambda_1 + \lambda_2) + G(\lambda_1 \lambda_2)]. \tag{2.4.35}$$

Finally, substituting Eqs. (2.4.26) and (2.4.27) into Eq. (2.4.35) results in

$$\frac{\cos \theta}{\delta\alpha_1 d\alpha_1} = \frac{1}{ds\delta s} [E + F \frac{(LG - NE)}{MG - NF} + G \frac{(LF - ME)}{MG - NF}]$$

$$= \frac{1}{ds\delta s} [\frac{EMG - ENF - FLG + ENF + FLG - EMG}{MG - NF}] = 0.$$

Thus, $\theta = 90^\circ$, and the principal directions are orthogonal. Since the principal directions are orthogonal we can choose the parametric curves to coincide with the directions of principal curvature. This provides additional simplification.

The equations of the lines of curvature are

$$\lambda_1 = \lambda_2 = 0. \tag{2.4.36}$$

If the lines of principal curvature coincide with the parametric lines, then from Eqs. (2.4.36), and (2.4.27)

$$LF - ME = 0. \tag{2.4.37}$$

From orthogonality, F = 0. This means, from Eq. (2.4.37), that

$$ME = 0. \tag{2.4.38}$$

For an actual surface, neither E nor G can be zero. Thus from Eq. (2.4.38), M = 0. Upon substituting F = M = 0 into Eqs. (2.4.21) nd (2.4.23), we find

$$k_1 = \frac{L}{E} \tag{2.4.39}$$

$$k_2 = \frac{N}{G}. \tag{2.4.40}$$

Equations (2.4.39) and (2.4.40) give the means for obtaining the principal curvature of a surface from the first and second fundamental quantities.

So far, we have considered space curves in general. The next step is to particularize the discussion to curves or surfaces of revolution. Following that we will deal with those surfaces of revolution of particular interest to mapping: the sphere or spheroid, and the plane, cone, or cylinder.

2.5 Surfaces of Revolution [10]

Surfaces of revolution are formed when a space curve is rotated about an axis. The two parameters needed to define a position on the surface of revolution will be z, and λ. Figure 2.5.1 gives the geometry for the development.

Let $R_0 = R_0(z)$. The position of point P in cartesian coordiantes, is

$$r = R_0 \cos \lambda \hat{i} + R_0 \sin \lambda \hat{j} + z\hat{k}. \tag{2.5.1}$$

Develop the vectors a_1 and a_2 at the point $P(z,\lambda)$, and the unit normal vector there. From Eqs. (2.2.2) and (2.2.3), we have

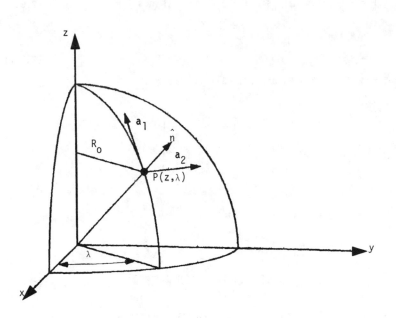

Figure 2.5.1   Geometry for a Surface of Revolution

$$\mathbf{a}_1 = \frac{\partial R_0}{\partial z} \cos \lambda \hat{i} + \frac{\partial R_0}{\partial z} \sin \lambda \hat{j} + \hat{k} \qquad (2.5.2)$$

$$\mathbf{a}_2 = - R_0 \sin \lambda \hat{i} + R_0 \cos \lambda \hat{j}. \qquad (2.5.3)$$

From Eqs. (2.3.8) and (2.3.11), the normal to the surface is

$$\hat{n} = - \frac{\mathbf{a}_1 \times \mathbf{a}_2}{EG - F^2}. \qquad (2.5.4)$$

It is now necessary to define the quantities of the first fundamental form in terms of z and λ. The first fundamental quantities are, from Eqs. (2.3.2), (2.5.2), and (2.5.3)

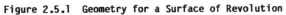

$$E = \left(\frac{\partial R_0}{\partial z}\right)^2 \cos^2\lambda + \left(\frac{\partial R_0}{\partial z}\right)^2 \sin^2\lambda + 1 = 1 + \left(\frac{\partial R_0}{\partial z}\right)^2 \qquad (2.5.5)$$

$$F = -\frac{\partial R_0}{\partial z}(\cos\lambda)(R_0\sin\lambda) + \frac{\partial R_0}{\partial z}(\sin\lambda)(R_0\cos\lambda) = 0 \qquad (2.5.6)$$

$$G = R_0^2\sin^2\lambda + R_0^2\cos^2\lambda$$

$$= R_0^2 \qquad (2.5.7)$$

Also, from Eqs. (2.5.2) and (2.5.3)

$$\mathbf{a}_1 \times \mathbf{a}_2 = \begin{vmatrix} \hat{i} & \hat{j} & \hat{k} \\ \frac{\partial R_0}{\partial z}\cos\lambda & \frac{\partial R_0}{\partial z}\sin\lambda & 1 \\ -R_0\sin\lambda & R_0\cos\lambda & 0 \end{vmatrix}$$

$$= -R_0\cos\lambda\hat{i} - R_0\sin\lambda\hat{j}$$

$$+ (R_0\frac{\partial R_0}{\partial z}\cos^2\lambda + R_0\frac{\partial R_0}{\partial z}\sin^2\lambda)\hat{k}$$

$$= -R_0(\cos\lambda\hat{i} + \sin\lambda\hat{j} - \frac{\partial R_0}{\partial z}\hat{k}). \qquad (2.5.8)$$

Substitute Eqs. (2.5.5), (2.5.6), (2.5.7), and (2.5.8) into Eq. (2.5.4), to obtain

$$\hat{n} = -\frac{R_0(\cos\lambda\hat{i} + \sin\lambda\hat{j} - \frac{\partial R_0}{\partial z}\hat{k})}{R_0\sqrt{1 + (\frac{\partial R_0}{\partial z})^2}}$$

$$= -\frac{\cos\lambda\hat{i} + \sin\lambda\hat{j} - \frac{\partial R_0}{\partial z}\hat{k}}{\sqrt{1 + (\frac{\partial R_0}{\partial z})^2}}. \qquad (2.5.9)$$

The quantities of the second fundamental form, in terms of z and λ, are given by Eqs. (2.4.16), (2.4.17), and (2.5.9), as

$$L = -\frac{\partial^2 \mathbf{r}}{\partial z^2} \cdot \hat{n}$$

$$= -(\frac{\partial^2 R_0}{\partial z^2}\cos\lambda\hat{i} + \frac{\partial^2 R_0}{\partial z^2}\sin\lambda\hat{j})$$

$$\cdot \frac{(\cos \lambda \hat{i} + \sin \lambda \hat{j} - \frac{\partial R_0}{\partial z} \hat{k})}{\sqrt{1 + (\frac{\partial R_0}{\partial z})^2}}$$

$$= -\frac{\frac{\partial^2 R_0}{\partial z^2} \cos^2 \lambda + \frac{\partial^2 R_0}{\partial z^2} \sin^2 \lambda}{\sqrt{1 + (\frac{\partial R_0}{\partial z})^2}}$$

$$= -\frac{\frac{\partial^2 R_0}{\partial z^2}}{\sqrt{1 + (\frac{\partial R_0}{\partial z})^2}} \cdot \qquad\qquad (2.5.10)$$

$$M = -\frac{\partial^2 \mathbf{r}}{\partial z \partial \lambda} \cdot \hat{n}$$

$$= -\left(-\frac{\partial R_0}{\partial z} \sin \lambda \hat{i} + \frac{\partial R_0}{\partial z} \cos \lambda \hat{j}\right)$$

$$\cdot \frac{\cos \lambda \hat{i} + \sin \lambda \hat{j} - (\frac{\partial R_0}{\partial z})\hat{k}}{\sqrt{1 + (\frac{\partial R_0}{\partial z})^2}}$$

$$= -\frac{(-\frac{\partial R_0}{\partial z} \sin \lambda \cos \lambda + \frac{\partial R_0}{\partial z} \sin \lambda \cos \lambda)}{\sqrt{1 + (\frac{\partial R_0}{\partial z})^2}} = 0 \qquad\qquad (2.5.11)$$

$$N = -\frac{\partial^2 \mathbf{r}}{\partial \lambda^2} \cdot \hat{n}$$

$$= -\left(-R_0 \cos \lambda \hat{i} - R_0 \sin \lambda \hat{j}\right)$$

$$\cdot \frac{(\cos \lambda \hat{i} + \sin \lambda \hat{j} - \frac{\partial R_0}{\partial z} \hat{k})}{\sqrt{1 + (\frac{\partial R_0}{\partial z})^2}}$$

$$= \frac{R_o \cos^2\lambda + R_o \sin^2\lambda}{\sqrt{1 + (\frac{\partial R_o}{\partial z})^2}}$$

$$N = \frac{R_o}{\sqrt{1 + (\frac{\partial R_o}{\partial z})^2}}. \qquad\qquad (2.5.12)$$

Note that both F and M are zero. Thus, our choice of parameters has gained us both orthogonality and coincidence with the principal directions.

The remaining quantities of the first and second fundamental forms give the curvatures in the principal directions. To obtain the curvature, substitute Eqs. (2.5.5), (2.5.7), (2.5.10) and (2.5.12) into Eqs. (2.4.39) and (2.4.40); this gives

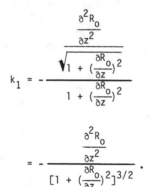

$$k_1 = -\frac{\dfrac{\frac{\partial^2 R_o}{\partial z^2}}{\sqrt{1 + (\frac{\partial R_o}{\partial z})^2}}}{1 + (\frac{\partial R_o}{\partial z})^2}$$

$$= -\frac{\frac{\partial^2 R_o}{\partial z^2}}{[1 + (\frac{\partial R_o}{\partial z})^2]^{3/2}}. \qquad\qquad (2.5.13)$$

This is the curvature in a plane perpendicular to the meridian. Note that this relation is comparable to Eq. (2.1.34).

Next, the curvature in the plane of the meridian is developed; its reciprocal gives the radius of curvature in this meridianal plane.

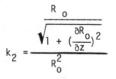

$$k_2 = \frac{\dfrac{R_o}{\sqrt{1 + (\frac{\partial R_o}{\partial z})^2}}}{R_o^2}$$

$$= \frac{1}{R_0 \sqrt{1 + (\frac{\partial R_0}{\partial z})^2}} \qquad (2.5.14)$$

R_2 is obtained from geometry, as developed below. From Figure 2.5.2, we can find the relations in the meridian plane. The colatitude is z.

$$\cos \phi = \frac{a_1}{\sqrt{E}} \cdot \hat{k}. \qquad (2.5.15)$$

Substitute Eqs. (2.5.2) and (2.5.5) into Eq. (2.5.15)

$$\cos \phi = \frac{(\frac{\partial R_0}{\partial z} \cos \lambda \hat{i} + \frac{\partial R_0}{\partial z} \sin \lambda \hat{j} + \hat{k})}{\sqrt{1 + (\frac{\partial R_0}{\partial z})^2}} \cdot \hat{k}$$

$$= \frac{1}{\sqrt{1 + (\frac{\partial R_0}{\partial z})^2}} \qquad (2.5.16)$$

From the figure

$$R_2 = \frac{R_0}{\cos \phi} \qquad (2.5.17)$$

Eliminating $\cos \phi$ between Eqs. (2.5.16) and (2.5.17)

$$R_2 = R_0 \sqrt{1 + (\frac{\partial R_0}{\partial z})^2} \qquad (2.5.18)$$

R_2 is the second radius of curvature, and is the reciprocal of Eq. (2.5.14).

2.6 Developable Surfaces [10]

We mentioned in Chapter 1 that there are two types of surfaces of interest to map projections: developable and non-developable. One way to make the distinction between the two is to consider the principal radii of

curvature.   Non-developable surfaces have two finite radii of curvature.
Developable surfaces have one finite and one infinite radius of curvature.
This section expands on the differences between the two types of surfaces.
The two developable ~~of~~ surfaces of interest to mapping, the cone and cylinder,
and the non-developable surface the sphere are considered.

The surfaces which are envelopes of one-parameter families of planes are
called developable surfaces. Every cone or cylinder is an envelope of a one-
parameter family of tangent planes. Moreover, every tangent plane has a

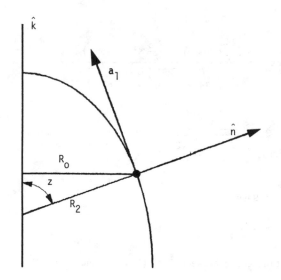

Figure 2.5.2 Geometry of the Meridian Curve

contact with the surface along a straight line. Consequently, a developable
surface is swept out by a family of rectilinear generators.

It is necessary, in the present section, to consider the tangent planes
for cones, cylinders, and spheres, and note their characteristics.   This
requires the investigation of the tangent plane defined by a_1 and a_2.

For the cone, consider arbitrary parameters u and v. Let the origin of the coordinate system be at the vertex of the cone. The parametric equation of the cone is

$$\mathbf{r} = v\mathbf{q}(u). \qquad (2.6.1)$$

From Eqs. (2.2.2) and (2.2.3)

$$\mathbf{a}_1 = v\dot{\mathbf{q}}(u) \qquad (2.6.2)$$

$$\mathbf{a}_2 = \mathbf{q}(u). \qquad (2.6.3)$$

Take the cross product of Eqs. (2.6.2) and (2.6.3)

$$\mathbf{a}_1 \times \mathbf{a}_2 = v\dot{\mathbf{q}}(u) \times \mathbf{q}(u). \qquad (2.6.4)$$

If $\mathbf{q}(u)$ and $\dot{\mathbf{q}}(u)$ are not collinear, the point (u,v) is regular, and the tangent plane has the following equation, after the substitution of Eqs. (2.6.1) and (2.6.4)

$$[\mathbf{r} - v\mathbf{q}(u)] \cdot v\mathbf{q}(u) \times \mathbf{q}(u) = 0$$

$$\mathbf{r} = \dot{\mathbf{q}}(u) \times \mathbf{q}(u) = 0. \qquad (2.6.5)$$

Thus, $\mathbf{r}$ from Eq. (2.6.5) depends only on u, and the family of tangent planes is a one-parameter family.

For a cylinder with elements parallel to a constant vector $\mathbf{c}$, the parametric equation is

$$\mathbf{r} = \mathbf{q}(u) + v\mathbf{c}. \qquad (2.6.6)$$

Applying Eqs. (2.2.2) and (2.2.3) to Eq. (2.2.6),

$$\mathbf{a}_1 = \dot{\mathbf{q}}(u) \qquad (2.6.7)$$

$$\mathbf{a}_2 = \mathbf{c}. \qquad (2.6.8)$$

Taking the cross product of Eqs. (2.6.7) and (2.6.8),

$$\mathbf{a}_1 \times \mathbf{a}_2 = \dot{\mathbf{q}}(u) \times \mathbf{c}. \qquad (2.6.9)$$

From Eqs. (2.6.6) and (2.6.9), the equation of the tangent plane is

$$[\mathbf{r} - \mathbf{q}(u) - v\mathbf{c}] \cdot \mathbf{q}(u) \times \mathbf{c} = 0$$

$$\mathbf{r} \cdot \mathbf{q}(u) \times \mathbf{c} = \mathbf{q}(u) \cdot \dot{\mathbf{q}}(u) \times \mathbf{c}. \qquad (2.6.10)$$

Again, we have an equation which depends only on the parameter u.

A different situation occurs when a non-developable surface such as the sphere is considered. Let the two parameters be ϕ and λ. The equation of the surface is

$$\mathbf{r} = a(\cos \lambda \cos \phi \hat{i} + \sin \lambda \cos \phi \hat{j} + \sin \phi \hat{k}). \qquad (2.6.11)$$

Using Eqs. (2.2.2) and (2.2.3),

$$\mathbf{a}_1 = a(-\cos \lambda \sin \phi \hat{i} - \sin \lambda \sin \phi \hat{j} + \cos \phi \hat{k}). \qquad (2.6.12)$$

$$\mathbf{a}_2 = a(-\sin \lambda \cos \phi \hat{i} + \cos \lambda \sin \phi \hat{j}) \qquad (2.6.13)$$

Taking the cross product of Eqs. (2.6.12) and (2.6.13)

$$\mathbf{a}_1 \times \mathbf{a}_2 = a^2 \begin{vmatrix} \hat{i} & \hat{j} & \hat{k} \\ -\cos \lambda \sin \phi & -\sin \lambda \cos \phi & \cos \phi \\ -\sin \lambda \cos \phi & \cos \lambda \cos \phi & 0 \end{vmatrix}$$

$$= a^2[-\cos \lambda \cos^2 \phi \hat{i} - \sin \lambda \cos^2 \phi \hat{j}$$

$$-\hat{k}(\cos^2 \lambda \sin \phi \cos \phi + \sin^2 \lambda \sin \phi \cos \phi)]$$

$$= -a^2[\cos \lambda \cos^2 \phi \hat{i} + \sin \lambda \cos^2 \phi \hat{j} + \cos \phi \sin \phi \hat{k}]. \qquad (2.6.14)$$

From Eqs. (2.6.11) and (2.6.14), the tangent plane to the sphere has the equation

$$[\mathbf{r} - a(\cos \lambda \cos \phi \hat{i} + \sin \lambda \cos \phi \hat{j} + \sin \phi \hat{k})]$$

$$[-a^2(\cos \lambda \cos^2 \phi \hat{i} + \sin \lambda \cos^2 \phi \hat{j} + \cos \phi \sin \phi \hat{k}] = 0. \qquad (2.6.15)$$

Equation (2.6.15) depends on two parameters, and thus, the sphere is a non-developable surface. Thus, the basic criterion for a developable surface is that the tangent plane must depend on only one parameter.

2.7 Transformation Matrix [19], [22]

A transformation matrix is derived below which permits the transformation from positions on the earth to places on the map. This entails relating the fundamental quantities of the earth and the plotting surfaces by means of a Jacobian determinant.

Consider the earth surface, with parametric curves on it defined
by ϕ and λ. The fundamental quantities will be defined as e, f, and g. The
coordinates of point P on the earth, as in Figure 2.7.1, are given
functionally as

$$\left.\begin{array}{l} x = x(\phi,\lambda) \\ y = y(\phi,\lambda) \\ z = z(\phi,\lambda) \end{array}\right\} . \qquad (2.7.1)$$

Consider next an arbitrary projection surface, with parametric curves
defined by the parameters u and v, with the fundamental quantities E', F' and
G'. The position of the point P' on the plotting surface in Figure 2.7.2 is
given functionally by

$$\left.\begin{array}{l} X = X(u,v) \\ Y = Y(u,v) \\ Z = Z(u,v) \end{array}\right\} . \qquad (2.7.2)$$

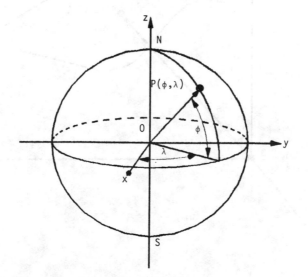

Figure 2.7.1 Parametric Representation of Point P of the Earth

The parametric curves on the earth are related to those on the projection surface by

$$u = u(\phi,\lambda) \atop v = v(\phi,\lambda) \Bigg\} \quad . \tag{2.7.3}$$

For the earth, and for any plotting surface, only two conditions are to be satisfied. The projection must be (1) unique, and (2) reversible. A point on the earth must correspond to only one point on the map, and vice versa. This requires that

$$\phi = \phi(u,v) \atop \lambda = \lambda(u,v) \Bigg\} \quad . \tag{2.7.4}$$

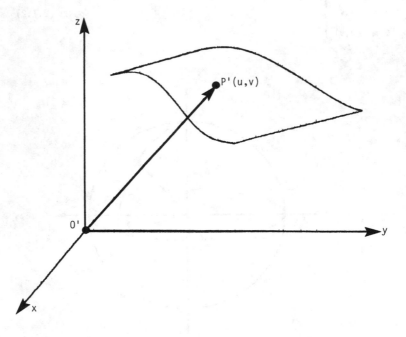

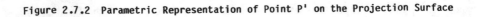

Figure 2.7.2 Parametric Representation of Point P' on the Projection Surface

Substituting Eq. (2.7.3) into Eq. (2.7.2), we have

$$\left.\begin{array}{l} X = X[u(\phi,\lambda),v(\phi,\lambda)] \\ Y = Y[u(\phi,\lambda),v(\phi,\lambda)] \\ Z = Z[u(\phi,\lambda),v(\phi,\lambda)] \end{array}\right\} . \qquad (2.7.5)$$

In this form, the surface will have the fundamental quantities E, F, and G.

Differentiate Eq. (2.7.5) with respect to ϕ, and λ; this gives

$$\left.\begin{array}{l} \dfrac{\partial X}{\partial \phi} = \dfrac{\partial X}{\partial u}\dfrac{\partial u}{\partial \phi} + \dfrac{\partial X}{\partial v}\dfrac{\partial v}{\partial \phi} \\[2mm] \dfrac{\partial X}{\partial \lambda} = \dfrac{\partial X}{\partial u}\dfrac{\partial u}{\partial \lambda} + \dfrac{\partial X}{\partial v}\dfrac{\partial v}{\partial \lambda} \\[2mm] \dfrac{\partial Y}{\partial \phi} = \dfrac{\partial Y}{\partial u}\dfrac{\partial u}{\partial \phi} + \dfrac{\partial Y}{\partial v}\dfrac{\partial v}{\partial \phi} \\[2mm] \dfrac{\partial Y}{\partial \lambda} = \dfrac{\partial Y}{\partial u}\dfrac{\partial u}{\partial \lambda} + \dfrac{\partial Y}{\partial v}\dfrac{\partial v}{\partial \lambda} \\[2mm] \dfrac{\partial Z}{\partial \phi} = \dfrac{\partial Z}{\partial u}\dfrac{\partial u}{\partial \phi} + \dfrac{\partial Z}{\partial v}\dfrac{\partial v}{\partial \phi} \\[2mm] \dfrac{\partial Z}{\partial \lambda} = \dfrac{\partial Z}{\partial u}\dfrac{\partial u}{\partial \lambda} + \dfrac{\partial Z}{\partial v}\dfrac{\partial v}{\partial \lambda} \end{array}\right\} . \qquad (2.7.6)$$

From Section 2.3, we know

$$\left.\begin{array}{l} E = (\dfrac{\partial X}{\partial \phi})^2 + (\dfrac{\partial Y}{\partial \phi})^2 + (\dfrac{\partial Z}{\partial \phi})^2 \\[2mm] F = \dfrac{\partial X}{\partial \phi}\dfrac{\partial X}{\partial \lambda} + \dfrac{\partial Y}{\partial \phi}\dfrac{\partial Y}{\partial \lambda} + \dfrac{\partial Z}{\partial \phi}\dfrac{\partial Z}{\partial \lambda} \\[2mm] G = (\dfrac{\partial X}{\partial \lambda})^2 + (\dfrac{\partial Y}{\partial \lambda})^2 + (\dfrac{\partial Z}{\partial \lambda})^2 \end{array}\right\} . \qquad (2.7.7)$$

Upon substituting Eq. (2.7.6) into Eq. (2.7.7), these become

$$E = (\dfrac{\partial X}{\partial u}\dfrac{\partial u}{\partial \phi})^2 + 2\,\dfrac{\partial X}{\partial u}\dfrac{\partial u}{\partial \phi}\dfrac{\partial X}{\partial v}\dfrac{\partial v}{\partial \phi} + (\dfrac{\partial X}{\partial v}\dfrac{\partial v}{\partial \phi})^2$$

$$+ (\dfrac{\partial Y}{\partial u}\dfrac{\partial u}{\partial \phi})^2 + 2\,\dfrac{\partial Y}{\partial u}\dfrac{\partial u}{\partial \phi}\dfrac{\partial X}{\partial v}\dfrac{\partial v}{\partial \phi} + (\dfrac{\partial Y}{\partial v}\dfrac{\partial v}{\partial \phi})^2$$

$$+ (\dfrac{\partial Z}{\partial u}\dfrac{\partial u}{\partial \phi})^2 + 2\,\dfrac{\partial Z}{\partial u}\dfrac{\partial u}{\partial \phi}\dfrac{\partial Z}{\partial v}\dfrac{\partial v}{\partial \phi} + (\dfrac{\partial Z}{\partial v}\dfrac{\partial v}{\partial \phi})^2$$

$$G = (\dfrac{\partial X}{\partial u}\dfrac{\partial u}{\partial \lambda})^2 + 2\,\dfrac{\partial X}{\partial u}\dfrac{\partial u}{\partial \lambda}\dfrac{\partial X}{\partial v}\dfrac{\partial v}{\partial \lambda} + (\dfrac{\partial X}{\partial v}\dfrac{\partial v}{\partial \lambda})^2$$

$$+ \left(\frac{\partial Y}{\partial u}\frac{\partial u}{\partial \lambda}\right)^2 + 2\,\frac{\partial Y}{\partial u}\frac{\partial u}{\partial \lambda}\frac{\partial Y}{\partial v}\frac{\partial v}{\partial \lambda} + \left(\frac{\partial Y}{\partial v}\frac{\partial v}{\partial \lambda}\right)^2$$

$$+ \left(\frac{\partial Z}{\partial u}\frac{\partial u}{\partial \lambda}\right)^2 + 2\,\frac{\partial Z}{\partial u}\frac{\partial u}{\partial \lambda}\frac{\partial Z}{\partial v}\frac{\partial v}{\partial \lambda} + \left(\frac{\partial Z}{\partial v}\frac{\partial v}{\partial \lambda}\right)^2$$

$$F = \left(\frac{\partial X}{\partial u}\right)^2 \frac{\partial u}{\partial \phi}\frac{\partial u}{\partial \lambda} + \frac{\partial X}{\partial u}\frac{\partial X}{\partial v}\left(\frac{\partial u}{\partial \phi}\frac{\partial v}{\partial \lambda} + \frac{\partial u}{\partial \lambda}\frac{\partial v}{\partial \phi}\right) + \left(\frac{\partial X}{\partial v}\right)^2 \frac{\partial v}{\partial \phi}\frac{\partial v}{\partial \lambda}$$

$$+ \left(\frac{\partial Y}{\partial u}\right)^2 \frac{\partial u}{\partial \phi}\frac{\partial u}{\partial \lambda} + \frac{\partial Y}{\partial u}\frac{\partial Y}{\partial v}\left(\frac{\partial u}{\partial \phi}\frac{\partial v}{\partial \lambda} + \frac{\partial u}{\partial \lambda}\frac{\partial v}{\partial \phi}\right) + \left(\frac{\partial Y}{\partial v}\right)^2 \frac{\partial v}{\partial \phi}\frac{\partial v}{\partial \lambda}$$

$$+ \left(\frac{\partial Z}{\partial u}\right)^2 \frac{\partial u}{\partial \phi}\frac{\partial u}{\partial \lambda} + \frac{\partial Z}{\partial u}\frac{\partial Z}{\partial v}\left(\frac{\partial u}{\partial \phi}\frac{\partial v}{\partial \lambda} + \frac{\partial u}{\partial \lambda}\frac{\partial v}{\partial \phi}\right) + \left(\frac{\partial Z}{\partial v}\right)^2 \frac{\partial v}{\partial \phi}\frac{\partial v}{\partial \lambda}. \quad (2.7.8)$$

Also, from Section 2.3, we have

$$E' = \left(\frac{\partial X}{\partial u}\right)^2 + \left(\frac{\partial Y}{\partial u}\right)^2 + \left(\frac{\partial Z}{\partial u}\right)^2$$

$$F' = \frac{\partial X}{\partial u}\frac{\partial X}{\partial v} + \frac{\partial Y}{\partial u}\frac{\partial Y}{\partial v} + \frac{\partial Z}{\partial u}\frac{\partial Z}{\partial v} \qquad \Bigg\} \;.$$

$$G' = \left(\frac{\partial X}{\partial v}\right)^2 + \left(\frac{\partial Y}{\partial v}\right)^2 + \left(\frac{\partial Z}{\partial v}\right)^2 \qquad\qquad\qquad (2.7.9)$$

Substituting Eq. (2.7.9) into Eq. (2.7.8), we find

$$E = \left(\frac{\partial u}{\partial \phi}\right)^2 E' + 2\,\frac{\partial u}{\partial \phi}\frac{\partial v}{\partial \phi} F + \left(\frac{\partial v}{\partial \phi}\right)^2 G'$$

$$F = \left(\frac{\partial u}{\partial \phi}\frac{\partial u}{\partial \lambda}\right) E' + \left(\frac{\partial u}{\partial \phi}\frac{\partial v}{\partial \lambda} + \frac{\partial u}{\partial \lambda}\frac{\partial v}{\partial \phi}\right) F' + \left(\frac{\partial v}{\partial \phi}\frac{\partial v}{\partial \lambda}\right) G'$$

$$G = \left(\frac{\partial u}{\partial \lambda}\right)^2 E' + 2\,\frac{\partial u}{\partial \lambda}\frac{\partial v}{\partial \lambda} F' + \left(\frac{\partial v}{\partial \lambda}\right)^2 G'. \quad (2.7.10)$$

Equations (2.9.10) may be written in matrix notation as

$$\begin{Bmatrix} E \\ F \\ G \end{Bmatrix} = \begin{bmatrix} \left(\frac{\partial u}{\partial \phi}\right)^2 & 2\,\frac{\partial u}{\partial \phi}\frac{\partial v}{\partial \phi} & \left(\frac{\partial v}{\partial \phi}\right)^2 \\[2mm] \frac{\partial u}{\partial \phi}\frac{\partial u}{\partial \lambda} & \frac{\partial u}{\partial \phi}\frac{\partial v}{\partial \lambda} + \frac{\partial u}{\partial \lambda}\frac{\partial v}{\partial \phi} & \frac{\partial v}{\partial \phi}\frac{\partial v}{\partial \lambda} \\[2mm] \left(\frac{\partial u}{\partial \lambda}\right)^2 & 2\,\frac{\partial u}{\partial \lambda}\frac{\partial v}{\partial \lambda} & \left(\frac{\partial v}{\partial \lambda}\right)^2 \end{bmatrix} \begin{Bmatrix} E' \\ F' \\ G' \end{Bmatrix} \quad (2.7.11)$$

The transformation matrix in Eq. (2.7.11) is the fundamental matrix for mapping transformations. This form of the transformation relation is of particular use in Chapter 5.

To facilitate the derivations of the following chapters it will be useful to develop the term

$$H = EG - F^2.$$ (2.7.12)

From Eq. (2.7.10)

$$H = [(\frac{\partial u}{\partial \phi})^2 E' + 2 \frac{\partial u}{\partial \phi} \frac{\partial v}{\partial \phi} F' + (\frac{\partial v}{\partial \phi})^2 G']$$

$$\times [(\frac{\partial u}{\partial \lambda})^2 E' + 2 \frac{\partial u}{\partial \lambda} \frac{\partial v}{\partial \lambda} F' + (\frac{\partial v}{\partial \lambda})^2 G']$$

$$- [\frac{\partial u}{\partial \phi} \frac{\partial u}{\partial \lambda} E' + (\frac{\partial u}{\partial \phi} \frac{\partial v}{\partial \lambda} + \frac{\partial u}{\partial \lambda} \frac{\partial v}{\partial \phi}) F' + \frac{\partial v}{\partial \phi} \frac{\partial v}{\partial \lambda} G']^2$$

$$H = (E')^2 (\frac{\partial u}{\partial \phi})^2 (\frac{\partial u}{\partial \lambda})^2 + 2 \frac{\partial u}{\partial \phi} \frac{\partial v}{\partial \phi} (\frac{\partial u}{\partial \lambda})^2 E'F'$$

$$+ (\frac{\partial v}{\partial \phi})^2 (\frac{\partial u}{\partial \lambda})^2 E'G' + 2 \frac{\partial u}{\partial \lambda} \frac{\partial v}{\partial \lambda} (\frac{\partial u}{\partial \phi})^2 E'F'$$

$$+ 4 \frac{\partial u}{\partial \phi} \frac{\partial v}{\partial \phi} \frac{\partial u}{\partial \lambda} \frac{\partial v}{\partial \lambda} (F')^2 + 2 \frac{\partial u}{\partial \lambda} \frac{\partial v}{\partial \lambda} (\frac{\partial v}{\partial \phi})^2 F'G'$$

$$+ (\frac{\partial u}{\partial \phi})^2 (\frac{\partial v}{\partial \lambda})^2 E'G' + 2 \frac{\partial u}{\partial \phi} \frac{\partial v}{\partial \phi} (\frac{\partial v}{\partial \lambda})^2 E'G'$$

$$+ (\frac{\partial u}{\partial \phi})^2 (\frac{\partial v}{\partial \lambda})^2 (G')^2$$

$$- (\frac{\partial u}{\partial \phi} \frac{\partial u}{\partial \phi})^2 (E')^2 - 2 (\frac{\partial u}{\partial \phi} \frac{\partial v}{\partial \lambda} + \frac{\partial u}{\partial \lambda} \frac{\partial v}{\partial \phi}) \frac{\partial u}{\partial \phi} \frac{\partial u}{\partial \lambda} E'F'$$

$$- 2 \frac{\partial u}{\partial \phi} \frac{\partial u}{\partial \lambda} \frac{\partial v}{\partial \phi} \frac{\partial v}{\partial \lambda} E'G' - (\frac{\partial u}{\partial \phi} \frac{\partial v}{\partial \lambda} + \frac{\partial u}{\partial x} \frac{\partial v}{\partial \phi})^2 (F')^2$$

$$- (\frac{\partial v}{\partial \phi} \frac{\partial v}{\partial \lambda})^2 (G')^2 - 2 (\frac{\partial u}{\partial \phi} \frac{\partial v}{\partial \lambda} + \frac{\partial u}{\partial \lambda} \frac{\partial v}{\partial \phi}) \frac{\partial v}{\partial \phi} \frac{\partial v}{\partial \lambda} F'G'$$

$$= E'G' [(\frac{\partial u}{\partial \phi} \frac{\partial v}{\partial \lambda})^2 + (\frac{\partial v}{\partial \phi} \frac{\partial u}{\partial \lambda})^2 - 2 (\frac{\partial u}{\partial \phi} \frac{\partial v}{\partial \lambda} \frac{\partial v}{\partial \phi} \frac{\partial u}{\partial \lambda})]$$

$$+ [4 \frac{\partial u}{\partial \phi} \frac{\partial u}{\partial \lambda} \frac{\partial v}{\partial \phi} \frac{\partial v}{\partial \lambda} - (\frac{\partial u}{\partial \phi} \frac{\partial v}{\partial \lambda} + \frac{\partial u}{\partial \lambda} \frac{\partial v}{\partial \phi})^2](F')^2$$

$$= E'G'[(\frac{\partial u}{\partial \phi} \frac{\partial v}{\partial \lambda})^2 + (\frac{\partial v}{\partial \phi} \frac{\partial u}{\partial \lambda})^2 - 2 (\frac{\partial u}{\partial \phi} \frac{\partial v}{\partial \lambda} \frac{\partial v}{\partial \phi} \frac{\partial u}{\partial \lambda})]$$

$$+ (F')^2 [4 \frac{\partial u}{\partial \phi} \frac{\partial u}{\partial \lambda} \frac{\partial v}{\partial \phi} \frac{\partial v}{\partial \phi} - (\frac{\partial u}{\partial \phi} \frac{\partial v}{\partial \lambda})^2$$

$$- (\frac{\partial u}{\partial \lambda} \frac{\partial v}{\partial \phi})^2 - 2 \frac{\partial u}{\partial \phi} \frac{\partial u}{\partial \lambda} \frac{\partial v}{\partial \phi} \frac{\partial v}{\partial \lambda}]$$

$$= [E'G' - (F')^2] \times [(\frac{\partial u}{\partial \phi} \frac{\partial v}{\partial \lambda})^2 + (\frac{\partial v}{\partial \phi} \frac{\partial u}{\partial \lambda})^2$$

$$- 2(\frac{\partial u}{\partial \phi} \frac{\partial v}{\partial \lambda} \frac{\partial v}{\partial \phi} \frac{\partial u}{\partial \lambda})]$$

$$EG - F^2 = [E'G' - (F')^2](\frac{\partial u}{\partial \phi} \frac{\partial v}{\partial \lambda} - \frac{\partial v}{\partial \phi} \frac{\partial u}{\partial \lambda})^2$$

$$= \begin{vmatrix} E' & F' \\ F' & G' \end{vmatrix} \begin{vmatrix} \frac{\partial u}{\partial \phi} & \frac{\partial u}{\partial \lambda} \\ \frac{\partial v}{\partial \phi} & \frac{\partial v}{\partial \lambda} \end{vmatrix}^2 . \tag{2.7.13}$$

The determinant

$$\begin{vmatrix} \frac{\partial u}{\partial \phi} & \frac{\partial u}{\partial \lambda} \\ \frac{\partial v}{\partial \phi} & \frac{\partial v}{\partial \lambda} \end{vmatrix}$$

is the Jacobian determinant of the transformation from (ϕ, λ) to (u, v).

A further simplification will be introduced, since we will be dealing with orthogonal curves: $f = F = F' = 0$. Substituting this into Eq. (2.7.13), we obtain

$$EG = E'G' \begin{vmatrix} \frac{\partial u}{\partial \phi} & \frac{\partial u}{\partial \lambda} \\ \frac{\partial u}{\partial \phi} & \frac{\partial v}{\partial \lambda} \end{vmatrix}^2 \tag{2.7.14}$$

This form will be of particular use in Chapter 4.

2.8 Conditions for Equal Area and Conformal Projections [6], [10]

The first fundamental form and the fundamental quantities are used to define the conditions for equal area and conformal projections. This can be done in a general manner. For the conventional projections, each case has its own requirements, and no general relations cna be defined. In the relations

that follow, e, g, and f refer to the model of the earth, and E, G, and F
refer to the plotting surface.

An equal area map is one in which the areas of domains are preserved as
they are transformed from the earth to the map. A theorem of differential
geometry requires that a mapping from the earth to the plotting surface is
locally equal area if, and only if,

$$eg - f^2 = EG - F^2.$$

In the orthogonal case, $f = F = 0$, and

$$eg = EG \qquad\qquad\qquad (2.8.1)$$

The relation (2.8.1) is substituted into Eq. (2.7.4) to obtain the equal
area transformation. This is done in Chapter 4 to transform from the earth to
the cylinder, plane, and cone.

A mapping of the surface of the earth onto the plane, or a developable
surface is called conformal (or isogonal) if it preserves the angle between
intersecting curves on the surface. From a theorem of differential geometry,
a mapping is called conformal if, and only if, the first fundamental forms of
the earth and the mapping surface, in compatible coordinates, are proportional
at every point. This requires that, for the orthogonal case,

$$\frac{E}{e} = \frac{G}{g} \qquad\qquad\qquad (2.8.2)$$

in the symbols of the previous section.

The transformation of Chapter 5 will apply these relations between the
earth, and the plane, cylinder, and cone.

2.9 Convergency of the Meridians [7], [18]

As one goes pole-ward form the equator on the earth, the meridians
converge, until, at the pole, all meridians intersect. This section gives an
estimate of the degree of this convergency as a function of latitude. Both
angular and linear convergency are considered.

In Figure 2.9.1, the fact of convergence of the meridians is apparent in the representation of 24 mile tracts in a township and range system. The appearance of convergency is a function of the latitude covered by the map. Between the tropic and the mid latitude, it is not too pronounced. The effect also depends on map scale.

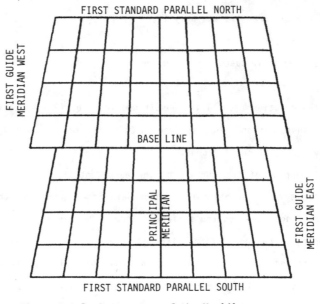

Figure 2.9.1 **Convergency of the Meridians**

We begin with angular convergence. In Figure 2.9.2 ACN and BDN are two meridians separated by a longitude difference of $\Delta\lambda$. Let CD be an arc of the circle of latitude ϕ. Let the earth be considered as spherical.

From the figure,

$$CD = CO'\Delta\lambda \tag{2.9.1}$$

$$DN = \frac{DO'}{\sin \phi}. \tag{2.9.2}$$

Approximately, the angle of convergency is

$$\theta = \frac{CD}{DN}. \tag{2.9.3}$$

Substituting Eqs. (2.9.1) and (2.9.2) into Eq. (2.9.3), and noting that CO' = DO'

$$\theta = \Delta\lambda \sin \phi. \qquad (2.9.4)$$

Let the distance between the meridians, measured along a parallel of latitude,

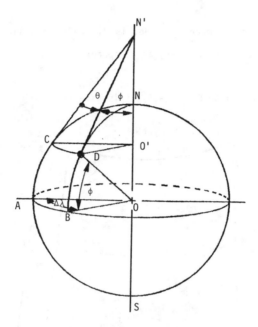

Figure 2.9.2 Geometry for Angular Convergence of the Meridians

be d, and let the radius of the earth be a. From the figure

$$\Delta\lambda = \frac{d}{a \cos \phi}. \qquad (2.9.5)$$

Substitute Eq. (2.9.5) into Eq. (2.9.4), to obtain

$$\theta = \frac{d \sin \phi}{a \cos \phi}$$

$$= \frac{d \tan \phi}{a}. \qquad (2.9.6)$$

θ is measured in radians.

The next step is to obtain the linear convergency. From Figure 2.9.3, let ℓ be the length of the meridian between two parallels ϕ_1 and ϕ_2. Let θ be the mean angular convergency at a mean latitude

$$\phi = \frac{\phi_1 + \phi_2}{2}.$$ (2.9.7)

The mean distance, at the mean latitude, is d. Define the linear convergency of the two meridians to be c. Then, as an approximation,

$$\theta = c/\ell.$$ (2.9.8)

Substitute Eq. (2.9.6) into Eq. (2.9.8), this gives

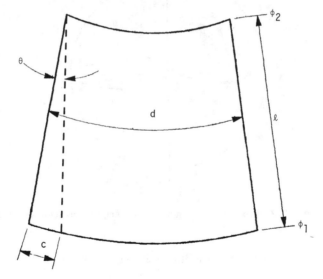

Figure 2.9.3 Geometry for Linear Convergence of the Meridians

$$c = \frac{d \cdot \ell \tan \phi}{a}.$$ (2.9.9)

2.10 Rotation of the Coordinate System [22]

A rotation of the coordinate system can be defined as a tool to obtain oblique, transverse, and equatorial projections from polar projections. This can be conveniently done by applying formulas of spherical trigonometry. The spherical trigonometry approach is justified, since in Chapters 4 and 5, it is shown that an intermediate transformation can be performed for the equal area and conformal projections which transforms from positions on the spheroidal earth to the authalic or conformal sphere, respectively. Once this is done, the rotation formulas for the sphere can be applied directly. Also, the conventional projections of Chapter 6 are based on a spherical earth for the majority of their practical applications.

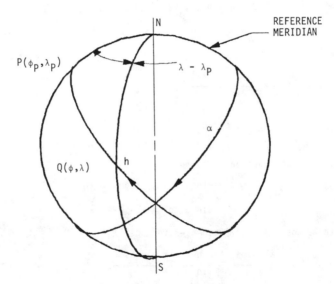

Figure 2.10.1 Geometry for the Rotational Transformation

Figure 2.10.1 provides the basic geometry for the rotational transformation. Let Q be any arbitrary point with coordinates ϕ and λ on the earth. Let P be the pole of the auxiliary spherical coordinate system. In the standard equatorial coordinate system, P has the coordinates ϕ_p and λ_p. Let h be the latitude of Q in the auxiliary system, and α, the longitude in that same system. A reference meridian is chosen for the origin of the measurement of α.

The intention is to derive the projection in the (h,α) system, and then transform to the (ϕ,λ) system for the plotting of the coordinates.

The relations between the angles of interest can be found from the spherical triangle PNQ.

From the law of cosines

$$\cos(90^0 - \phi) = \cos(90^0 - \phi_p)\cos(90^0 - h)$$
$$+ \sin(90^0 - \phi_p)\sin(90^0 - h)\cos \alpha$$
$$\sin \phi = \sin \phi_p \sin h + \cos \phi_p \cos h \cos \alpha. \qquad (2.10.1)$$

From the law of sines

$$\frac{\sin(\lambda - \lambda_p)}{\sin(90^0 - h)} = \frac{\sin \alpha}{\sin(90^0 - \phi)}$$

$$\sin(\lambda - \lambda_p) = \frac{\sin \alpha \cos h}{\cos \phi}. \qquad (2.10.2)$$

Also, applying the four parts formula

$$\cos(90^0 - \phi_p)\cos \alpha = \sin(90^0 - \phi_p)\cot(90^0 - h)$$
$$- \sin \alpha \cot(\lambda - \lambda_p)$$
$$\sin \phi_p \cos \alpha = \cos \phi_p \tan h - \sin \alpha \cot(\lambda - \lambda_p)$$
$$\cot(\lambda - \lambda_p) = \frac{\cos \phi_p \tan h - \sin \phi_p \cos \alpha}{\sin \alpha}. \qquad (2.10.3)$$

The inverse relationships are also of use. From the law of cosines

$$\cos(90^0 - h) = \cos(90^0 - \phi)\cos(90^0 - \phi_p)$$
$$+ \sin(90^0 - \phi)\sin(90^0 - \phi_p)\cos(\lambda - \lambda_p)$$
$$\sin h = \sin \phi \sin \phi_p + \cos \phi \cos \phi_p \cos (\lambda - \lambda_p). \qquad (2.10.4)$$

From the four parts formula

$$\cos(90^0 - \phi_p)\cos(\lambda - \lambda_p) = \sin(90^0 - \phi_p)\cot(90^0 - \phi)$$
$$- \sin(\lambda - \lambda_p)\cot \alpha$$

$$\sin \phi_p \cos(\lambda - \lambda_p) = \cos \phi_p \tan \phi - \sin(\lambda - \lambda_p)\cot \alpha$$

$$\sin(\lambda - \lambda_p)\cot \alpha = \cos \phi_p \tan \phi - \sin \phi_p \cos (\lambda - \lambda_p)$$

$$\tan \alpha = \frac{\sin (\lambda - \lambda_p)}{\cos \phi_p \tan \phi - \sin \phi_p \cos(\lambda - \lambda_p)}. \qquad (2.10.5)$$

A final useful equation is needed for unique quadrant determination. From a consideration of Figure 2.10.1

$$\cos \alpha \cos h = \sin \phi \cos \phi_p - \cos \phi \sin \phi_p \cos (\lambda - \lambda_p). \qquad (2.10.6)$$

Having possession of equations (2.10.4), (2.10.5), and (2.10.6), we can accomplish any rotations necessary to form oblique, transverse, and equatorial projections from polar and regular projections. These will be required in some of the projections of Chapters 4, 5, and 6.

PROBLEMS

2.1 Given the plane curve $Y = X^3 + 2X^2 + 3$. Find the curvature and radius of curvature at $X = 3$. What is the torsion?

2.2 Given $X = C_1 \lambda + C_2$, where C_1 and C_2 are constants, and $Y = Y(\phi)$. Determine the Jacobian determinant.

2.3 The first fundamental form for the sphere is $(ds)^2 = R^2 (d\phi)^2 + R^2 \cos^2\phi(d\lambda)^2$. Find the following:
a) The distance integral between the arbitrary points P_1 and P_2.
b) The incremental area.

2.4 Let the radius of a spherical earth be 6,378,100 meters, and the distance measured along the parallel be 2,000 meters. What is the angle of convergency of the meridians at a latitude of 45 N?

2.5 Given the data of Problem 2.4, where the distance along the parallel is the mean distance. Let the length of the meridian be 1,000 meters. What is the linear convergency of the meridians?

2.6 The coordinate of the pole of the auxiliary coordinate system are latitude 45° N, and longitude 0°. The coordinates of a geographic location in the auxiliary system are $\alpha = 30°$ and $h = 30°$. What are the coordinates in the standard equatorial system?

2.7 The coordinates of the pole of the auxiliary coordinate system are latitude 45° N, and longitude 0°. The standard equatorial coordinates are $\phi = 35^0$, and $\lambda = 30^0$. Find the corresponding coordinates in the auxiliary system.

3

Figure of the Earth

The basic geometrical surface taken as the model of the earth is an oblate spheroid generated by revolving an ellipse about its semi-minor axis. This chapter is be concerned with the geometry of the spheroid, and the reduction of the geometry to the spherical case.

The figure of the earth, as seen by the cartographer, is far less complex than that seen by the geodesist or the astronomer. For his basic surface, the cartographer may assume a single best spheroid, and project thereon positions from an undulating earth. Then, he is free to begin the process of projecting onto a map. The geodesist must consider a spheroid which is possibly best in one portion of the world, and other spheroids which are best in other portions of the world, and then strain to patch them together in a coordinated manner. The astronomer, dealing with the dynamical figure of the earth, must consider pear-shape and undulations to obtain solutions couched in tesseral harmonics.

The cartographer has only to deal with the geometrical figure of the earth, and can enjoy immense simplifications. To facilitate the transformations of Chapters 4, 5, and 6, it is necessary to consider the geometry of the ellipse and the spheroid. The coordinate system of the spheroid is be introduced. Angles and distances on the spheroid are considered. Then, particular constants for the actual size and shape of the earth are be given.

Many of the projections in Chapters 4, 5, and 6 are be based on an intermediate transformation to a sphere. Thus, it is necessary to investigate the further simplifications in coordinates, angles and distances on a sphere.

3.1 Geometry of the Ellipse [16], [17]

The ellipse is the generating curve which produces the spheroid of revolution. The nomenclature of the ellipse is best described with reference to Figure 3.1.1.

The semi-major axis, a, is the length of the line AO, or the line OB. The semi-minor axis, b, is the length of the line DO, or the line CO. The equation for the ellipse, for a cartesian coordinate system with origin at O, is

$$\frac{x^2}{a^2} + \frac{z^2}{b^2} = 1. \qquad (3.1.1)$$

The degree of departure from circularity is described by the eccentricity, e, or the flattening f. The eccentricity, flattening, semi-

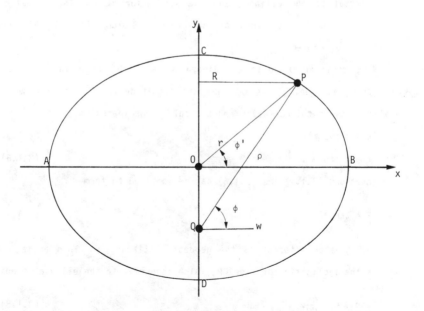

Figure 3.1.1 Geometry of the Ellipse

major axis, and semi-minor axis are related as follows

$$e^2 = \frac{a^2 - b^2}{a^2} \qquad\qquad\qquad (3.1.2)$$

$$f = \frac{a - b}{a} \qquad\qquad\qquad (3.1.3)$$

$$e^2 = 2f - f^2. \qquad\qquad\qquad (3.1.4)$$

At this point we introduce one of the two angular coordinates which uniquely locate a position on the spheroid. This first coordinate is the latitude. Two types of latitude will be noted: the geodetic and the geocentric. The relation between the two is now derived.

The geocentric latitude is the angle between a vector from a center of the ellipse, to a point P, on the ellipse, or meridian, and the semi-major axis. The geodetic latitude is the angle between a line through the given point, normal to the ellipse, and the semi-major axis. The normal to the ellipse is the line defined by a surveyor's plumb line if all gravity anomalies are ignored.

Now, consider a polar coordinate sysem with the origin at 0. The geocentric latitude, ϕ', is OP, and the magnitude of the vector is r. The relation between the cartesian and the polar coordinates is

$$x = r \cos \phi' \qquad\qquad\qquad (3.1.5)$$

$$z = r \sin \phi'. \qquad\qquad\qquad (3.1.6)$$

Equations (3.1.5) and (3.1.6) can be combined to form

$$\frac{z}{x} = \tan \phi'. \qquad\qquad\qquad (3.1.7)$$

Of greater interest is the geodetic latitude, ϕ. This angle, PQW, defines the inclination of line QP, which is normal to the ellipse at point P.

$$\tan \phi = -\frac{dx}{dz}. \qquad\qquad\qquad (3.1.8)$$

Taking the differential of Eq. (3.1.1)

$$\frac{2x\ dx}{a^2} + \frac{2z\ dz}{b^2} = 0$$

$$\frac{dx}{dz} = -\frac{a^2}{b^2}\frac{z}{x}.\qquad\qquad(3.1.9)$$

Substitute Eq. (3.1.9) into Eq. (3.1.8), to obtain

$$\tan\ \phi = \frac{a^2}{b^2}\frac{z}{x}.\qquad\qquad(3.1.10)$$

Substitute Eq. (3.1.7) into Eq. (3.1.10), this gives

$$\tan\ \phi = \frac{a^2}{b^2}\tan\ \phi'$$

$$\tan\ \phi' = \frac{b^2}{a^2}\tan\ \phi.\qquad\qquad(3.1.11)$$

Finally substitute Eq. (3.1.2) into Eq. (3.1.12), to arrive at

$$\tan\ \phi' = (1 - e^2)\tan\ \phi.\qquad\qquad(3.1.12)$$

Table 3.1.1
Geocentric and Geodetic Latitude for the
WGS-72 Spheroid (Degrees)

Geodetic ϕ	Geocentric ϕ'	Geocentric ϕ'	Geodetic ϕ
0.00	0.0000	0.00	0.0000
5.00	4.9667	5.00	5.0335
10.00	9.9344	10.00	10.0660
15.00	14.9041	15.00	15.0965
20.00	19.8766	20.00	20.1240
25.00	24.8529	25.00	25.1477
30.00	29.8337	30.00	30.1669
35.00	34.8194	35.00	35.1810
40.00	39.8106	40.00	40.1896
45.00	44.8076	45.00	45.1925
50.00	49.8104	50.00	50.1894
55.00	54.8190	55.00	55.1806
60.00	59.8331	60.00	60.1664
65.00	64.8523	65.00	65.1471
70.00	69.8761	70.00	70.1234
75.00	74.9036	75.00	75.0960
80.00	79.9340	80.00	80.0557
85.00	84.9665	85.00	85.0334
90.00	90.0000	90.00	90.0000

Table 3.1.1 gives the relation between geocentric latitude and geodetic latitude for the WGS-72 spheroid, which is be discussed in Section 3.4.

The convention for measuring geodetic latitude is $+\phi$ in the northern hemisphere and $-\phi$ in the southern hemisphere.

3.2 Geometry of the Spheroid [23]

The spheroid, which is taken as the model of the earth, is obtained by revolving the ellipse of Figure 3.1.1 about the z-axis. For the cartesian coordinate system shown in Figure 3.2.1, the equation of the spheroidal surface is

$$\frac{x^2}{a^2} + \frac{y^2}{a^2} + \frac{z^2}{b^2} = 1. \tag{3.2.1}$$

The nomenclature of the spheroid can be obtained by studying Figure 3.2.1. Each of the infinity of positions of the ellipse as it is rotated about the z-axis defines a meridinal ellipse, or meridian. The angle λ, meausred in the x-y plane, and from the x-axis, is the longitude of any and all points on the meridianal ellipse. This is the second of the two

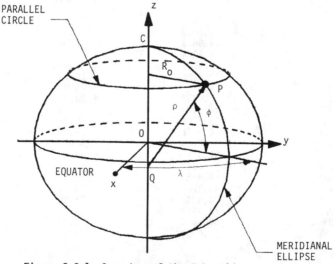

Figure 3.2.1 Geometry of the Spheroid

angular coordinates which uniquely define a position on the spheroid. As a convention, a rotation from $+x$ to $+y$, or east, will be positive, and the reverse rotation, negative.

Consider the point P in Figure 3.2.1 to be defined by ϕ and λ. Suppose now that λ is allowed to vary while ϕ is held constant. The locus on the spheroid traced out by P is a circle of parallel of radius R_0. The circle of parallel for a latitude of zero is the equator.

It remains to derive the equations for several radii of importance in future developments. These are the two principle radii of curvature, and the radius of a parallel circle, all as a function of latitude.

Consider the meridianal ellipse at any arbitrary λ. From Eq. (3.2.1)

$$\frac{R_0^2}{a^2} + \frac{z^2}{b^2} = 1 \tag{3.2.2}$$

where $R_0 = \sqrt{x^2 + y^2}$ is the radius of a parallel circle.

Substitute Eq. (3.1.2) into Eq. (3.2.2), to obtain

$$\frac{R_0^2}{a^2} + \frac{z^2}{a^2(1 - e^2)} = 1$$

$$R_0^2(1 - e^2) + z^2 = a^2(1 - e^2). \tag{3.2.3}$$

Taking the differential of Eq. (3.2.3) to obtain the slope of the tangent at P, we have

$$2R_0 \, dR_0(1 - e^2) + 2z \, dz = 0$$

$$\frac{dz}{dR_0} = -\frac{R_0}{a}(1 - e^2). \tag{3.2.4}$$

The slope of the normal at P is again

$$-\frac{dR_0}{dz} = \frac{z}{R_0(1 - e^2)} = \tan \phi$$

$$z = R_0(1 - e^2)\tan \phi. \tag{3.2.5}$$

From Figure 3.2.1

$$\sin \phi = \frac{z}{(1 - e^2)\overline{QP}}$$

$$z = \overline{QP}(1 - e^2)\sin \phi. \qquad (3.2.6)$$

Substitute Eq. (3.2.5) into Eq. (3.2.3); this gives

$$R_0^2(1 - e^2) + R_0^2(1 - e^2)^2\tan^2\phi = a^2(1 - e^2)$$

$$R_0^2 + R_0^2(1 - e^2)\tan^2\phi = a^2$$

$$(\cos^2\phi + \sin^2\phi)R_0^2 - e^2R_0^2\sin^2\phi = a^2\cos^2\phi$$

$$R_0^2(1 - e^2\sin^2\phi) = a^2\cos^2\phi$$

$$R_0 = \frac{a \cos \phi}{\sqrt{1 - e^2\sin^2\phi}}. \qquad (3.2.7)$$

Also, from Figure 3.2.1

$$R_0 = \overline{QR} \cos \phi. \qquad (3.2.8)$$

Equating Eqs. (3.2.7) and (3.2.8), we find

$$\overline{QP} \cos \phi = \frac{a \cos \phi}{\sqrt{1 - e^2\sin^2\phi}}$$

$$\overline{QP} = \frac{a}{\sqrt{1 - e^2\sin^2\phi}}.$$

$\overline{QP}$ is the radius of curvature of the spheroid in the plane perpendicular to the meridianal plane, and will be denoted as R_p.

$$R_p = \frac{a}{\sqrt{1 - e^2\sin^2\phi}}. \qquad (3.2.9)$$

The radius of curvature of the meridianal ellipse follows from the formula for a plane curve Eq. (2.1.35).

$$R_M = \left| \frac{[1 + (\frac{dz}{dR_0})^2]^{3/2}}{\frac{d^2z}{dR_0^2}} \right| . \tag{3.2.10}$$

From Eq. (3.2.4),

$$\frac{d^2z}{dR^2} = -\frac{1}{z}(1 - e^2) + \frac{R_0}{z^2}(1 - e^2)\frac{dz}{dR}. \tag{3.2.11}$$

Substituting Eq. (3.2.4) into Eq. (3.2.10), we have

$$\frac{d^2z}{dR_0^2} = \frac{1}{z}\left[-(1 - e^2) - (\frac{dz}{dR_0})^2\right]$$

$$= -\frac{1}{z}\left[1 + (\frac{dz}{dR_0})^2 - 3^2\right]. \tag{3.2.12}$$

Since the slope of the normal is tan ϕ, that of the tangent is $-\cot \phi$, then

$$\frac{dz}{dR_0} = \cot \phi. \tag{3.2.13}$$

Substituting Eq. (3.2.13) into Eq. (3.2.12), we obtain

$$\frac{dz}{dR_0} = -\frac{1 + \cot^2\phi - e^2}{z}$$

$$= \frac{\frac{1}{\sin^2\phi} - e^2}{z}. \tag{3.2.14}$$

From Eqs. (3.2.6) and (3.2.9), we have

$$z = \frac{a(1 - e^2)\sin \phi}{1 - e^2\sin^2\phi}. \tag{3.2.15}$$

Then, substituting Eqs. (3.2.13), (3.2.14), and (3.2.15) into Eq. (3.2.10),

$$R_m = \left| \frac{(1 + \frac{\cos^2\phi}{\sin^2\phi})^{3/2}}{-(\frac{1}{\sin^2\phi} - e^2)\frac{1 - e^2\sin^2\phi}{a(1 - e^2)\sin \phi}} \right|$$

$$= \left| \frac{(\frac{\cos^2\phi + \sin^2\phi}{\sin^2\phi})^{3/2}}{-(\frac{1 - e^2\sin^2\phi}{\sin^2\phi}) \frac{1 - e^2\sin^2\phi}{a(1 - e^2)\sin\phi}} \right|$$

$$= \left| -\frac{\frac{1}{\sin^3\phi}}{(\frac{1 - e^2\sin^2\phi}{\sin^3\phi})^{3/2} \cdot \frac{1}{a(1 - e^2)}} \right|$$

$$= \left| \frac{a(1 - e^2)}{(1 - e^2\sin^2\phi)^{3/2}} \right| . \qquad (3.2.16)$$

Equations (3.2.7), (3.2.9), and (3.2.16) give the radius of the circle of parallel, the radius of curvature normal to the meridian, and the radius of curvature in the meridianal plane, respectively. The radii of curvature and the radius of the circle of parallel are tabulated in Table 3.2.1 as functions of latitude for the WGS-72 spheroid (Section 3.5).

Before turning to distances on the spheroid, it will be useful to relate the cartesian coordinates for the spheroid to the polar coordinates.

$$\left. \begin{array}{l} x = R_p \cos\phi \cos\lambda \\ y = R_p \cos\phi \sin\lambda \\ z = (1 - e^2)R_p\sin\phi \end{array} \right\} . \qquad (3.2.17)$$

The first fundamental form of the spheroid (Section 2.3) becomes

$$(ds)^2 = R_m^2(d\phi)^2 + R_p^2\cos^2\phi(d\lambda)^2 . \qquad (3.2.18)$$

Table 3.2.1
Radii as a Function of Latitude for the
WGS-72 Spheroid

Geodetic Latitude ϕ	Radii of Curvature		Radius of Parallel Circle R_o
	R_p	R_m	
(Degrees)	(Meters)	(Meters)	(Meters)
0.00	6378135.	6335439.	6378135.
5.00	6378297.	6335922.	6354026.
10.00	6378779.	6337357.	6281871.
15.00	6379566.	6339702.	6162187.
20.00	6380634.	6342888.	5995834.
25.00	6381951.	6346818.	5784012.
30.00	6383479.	6351376.	5528254.
35.00	6385170.	6356425.	5230424.
40.00	6386974.	6361814.	4892705.
45.00	6388836.	6367380.	4517588.
50.00	6390700.	6372954.	4107861.
55.00	6392508.	6378366.	3666590.
60.00	6394207.	6383452.	3197101.
65.00	6395743.	6388054.	2702955.
70.00	6397070.	6392031.	2187924.
75.00	6398147.	6395260.	1655959.
80.00	6398941.	6397641.	1111161.
85.00	6399427.	6399100.	557743.
90.00	6399591.	6399591.	0.

Equation (3.2.18) is found to be useful in the transformations from the spheroid to the sphere in Chapters 4 and 5, and in the discussion of distortions in Chapter 7.

3.3 Distances and Angles on the Spheroid [5]

Three types of distances measured on the spheroid are be considered. These are distances along a circle of parallel, distances along the meridianal ellipse, and distances between two ordinary points.

We can deal with the distance along a circle of parallel very easily. From Eq. (3.2.7)

$$d = \frac{a\Delta\lambda \cos \phi}{\sqrt{1 - e^2\sin^2\phi}} \qquad (3.3.1)$$

where $\Delta\lambda$ is the angular separation of two points on the circle of parallel, in radians. Table 3.3.1 gives the distance for an angular separation of 1' as a function of latitude for the WGS-72 ellipsoid.

Distance along the meridianal ellipse requires an integration of Eq. (3.2.16). To facilitate this, Eq. (3.2.16) is expanded by the binomial theorem.

$$R_m = a(1 - e^2)(1 - \frac{3}{2} e^2 \sin^2 \phi + \frac{15}{8} e^4 \sin^4 \phi + \frac{35}{16} e^6 \sin^6 \phi + \ldots). \qquad (3.3.2)$$

Equation (3.3.2) is a rapidly converging series, as we shall see when we display the values of e in Section 3.4.

The distance between positions at latitude ϕ_1 and ϕ_2 on the same meridianal ellipse is given by:

$$d = \int_{\phi_1}^{\phi_2} R_m \, d\phi. \qquad (3.3.3)$$

Substitute Eq. (3.3.2) into Eq. (3.3.3) and integrate, to obtain

$$d = a(1 - e^2) \int_{\phi_1}^{\phi_2} (1 + \frac{3}{2} e^2 \sin^2 \phi + \frac{15}{8} e^4 \sin^4 \phi + \frac{35}{16} e^6 \sin \phi + \ldots) d\phi$$

$$d = a(1 - e^2) \{ \phi + \frac{3}{2} e^2 (\frac{\phi}{2} - \frac{1}{4} \sin 2\phi)$$

$$+ \frac{15}{8} e^4 (\frac{3\phi}{8} - \frac{\sin 2\phi}{4} + \frac{\sin 4\phi}{32})$$

$$+ \frac{35}{16} e^6 [- \frac{\sin^5 \phi \cos \phi}{6} + \frac{5}{6} [\frac{3\phi}{8} - \frac{\sin 2\phi}{4} + \frac{\sin 4\phi}{32}] + \ldots \}_{\phi_1}^{\phi_2}$$

$$= a(1 - e^2) \{ \phi(1 + \frac{3}{4} e^2 + \frac{45}{64} e^4 + \frac{525}{768} e^6 + \ldots)$$

$$- \sin 2\phi (\frac{3}{8} e^2 + \frac{15}{32} e^4 + \frac{175}{384} e^6 + \ldots)$$

$$+ \sin 4\phi \left(\frac{15}{156} e^4 + \frac{175}{3072} e^6 + \dots \right)$$

$$\left. - \frac{35}{96} e^6 \sin^5 \phi \cos \phi + \dots \right\}_{\phi_1}^{\phi_2}$$

$$= a \left\{ \left(1 - e^2 + \frac{3}{4} e^2 - \frac{3}{4} e^4 + \frac{45}{64} e^4 - \frac{45}{64} e^6 + \frac{525}{768} e^6 + \dots \right) \phi \right.$$

$$- \left(\frac{3}{8} e^2 - \frac{3}{8} e^4 + \frac{15}{32} e^4 - \frac{15}{32} e^6 + \frac{175}{384} e^6 + \dots \right) \sin^2 \phi$$

$$+ \left(\frac{15}{256} e^4 - \frac{15}{256} e^6 + \frac{175}{3072} e^8 + \dots \right) \sin 4\phi$$

$$\left. - \frac{35}{96} e^6 \left(\frac{\sin \phi \cos \phi}{8} \right) (3 - 4 \cos 2\phi + \cos 4\phi) \right._{\phi_1}^{\phi_2}$$

$$d = a \left\{ \left(1 - \frac{e^2}{4} - \frac{3}{64} e^4 - \frac{5}{256} e^6 - \dots \right) \phi \right.$$

$$- \left(\frac{3}{8} e^2 + \frac{3}{32} e^4 - \frac{5}{384} e^6 + \dots \right) \sin 2\phi$$

$$+ \left(\frac{15}{256} e^4 - \frac{5}{3072} e^6 + \dots \right) \sin 4\phi$$

$$\left. - \frac{35}{96} e^6 \left(\frac{\sin 2\phi}{16} \right) (3 - 4 \cos 2\phi + \cos 4\phi) \right\}_{\phi_1}^{\phi_2}. \qquad (3.3.4)$$

Further, expanding the last term in Eq. (3.3.4),

$$- \frac{35}{96} e^6 \left(\frac{\sin 2\phi}{16} \right) (3 - 4 \cos 2\phi + \cos 4\phi)$$

$$= - \frac{35}{96} e^6 \left(\frac{3}{16} \sin 2\phi - \frac{1}{4} \sin 2\phi \cos 2\phi + \frac{1}{16} \sin 2\phi \cos 4\phi \right)$$

$$= - \frac{35}{96} e^6 \left(\frac{3}{16} \sin 2\phi - \frac{1}{8} \sin 4\phi - \frac{1}{32} \sin 2\phi + \frac{1}{32} \sin 6\phi \right). \qquad (3.3.5)$$

Finally, substitute Eq. (3.3.5) into Eq. (3.3.4),

$$d = a\{(1 - \frac{e^2}{4} - \frac{3}{64}e^4 - \frac{5}{256}e^6 - \ldots)\phi$$

$$- [\frac{3}{8}e^2 + \frac{3}{32}e^4 + (-\frac{5}{384} + \frac{35}{96}(\frac{3}{16} - \frac{1}{32}))e^6 + \ldots]\sin 2\phi$$

$$+ [\frac{15}{256}e^4 + (-\frac{5}{3072} + \frac{35}{96} \cdot \frac{1}{8})e^6 + \ldots]\sin 4\phi$$

$$- \frac{35}{96} \cdot \frac{e^6}{32}\sin 6\phi + \ldots\}_{\phi_1}^{\phi_2}$$

$$d = a\{(1 - \frac{e^2}{4} - \frac{3}{64}e^4 - \frac{5}{256}e^6 - \ldots)\phi$$

$$- (\frac{3}{8}e^2 + \frac{3}{32}e^4 + \frac{45}{1024}e^6 + \ldots)\sin 2\phi \qquad (3.3.6)$$

$$+ (\frac{15}{256}e^4 + \frac{45}{1024}e^6 + \ldots)\sin 4\phi - \frac{35}{3072}e^6\sin 6\phi + \ldots\}_{\phi_1}^{\phi_2}$$

Equation (3.3.6) has been evaluated in Table 3.3.2 for 1' intervals of arc along the meridinial ellipse as a function of latitude for the WGS-72 spheroid.

The distance along a spheroid between two arbitrary points $P_1(\phi,\lambda_1)$ and $P_2(\phi_2,\lambda_2)$ is obtained from consideration of the first fundamental form for a spheroid (3.2.18).

$$s = \int_{\phi_1}^{\phi_2} [R_m^2 + R_p^2\cos^2\phi(\frac{d\lambda}{d\phi})^2]^{1/2}d\phi. \qquad (3.3.7)$$

To obtain the shortest line on the spheroid connecting P_1 and P_2, that is, the geodesic curve, apply the Euler-Lagrange minimization to Eq. (3.3.7), where

$$L(\phi,\lambda,\lambda') = R_m^2 + R_p^2\cos^2\phi(\lambda')^2$$

$$\frac{d}{d\phi}(\frac{\partial L}{\partial\lambda'}) = \frac{\partial L}{\partial\lambda} = 0$$

Table 3.3.1
Distances Along the Circle of Parallel for a Separation of 1', for the WGS-72 Spehroid.

Geodetic Latitude ϕ (Degrees)	Distance d (Meters)
0.00	1855.323
5.00	1848.310
10.00	1827.321
15.00	1792.506
20.00	1744.116
25.00	1682.500
30.00	1608.103
35.00	1521.468
40.00	1423.229
45.00	1314.112
50.00	1194.927
55.00	1066.567
60.00	929.998
65.00	786.257
70.00	636.441
75.00	481.699
80.00	323.223
85.00	162.241
90.00	0.000

Table 3.3.2
Distance Along the Meridianal Ellipse for a Separation of 1', for WG5-72 Spheroid

Geodetic Latitude ϕ (Degrees)	Distance d (Meters)
0.00	1842.903
5.00	1843.044
10.00	1843.461
15.00	1844.143
20.00	1845.070
25.00	1846.213
30.00	1847.539
35.00	1849.008
40.00	1850.575
45.00	1852.194
50.00	1853.816
55.00	1855.390
60.00	1856.870
65.00	1858.208
70.00	1859.365
75.00	1860.304
80.00	1860.997
85.00	1861.421
90.00	1861.564

$$\frac{\partial L}{\partial \lambda} = R_p^2 \cos^2 \phi \frac{d\lambda}{d\phi} = c \tag{3.3.8}$$

where c is a constant.

Substitute Eq. (3.3.8) into Eq. (3.3.7), to obtain

$$s = \int_{\phi_1}^{\phi_2} (R_m^2 + \frac{c^2}{R_p^2 \cos^2 \phi})^{1/2} d\phi. \tag{3.3.9}$$

It remains to evaluate c in Eq. (3.3.9). Integrate Eq. (3.3.8); this gives

$$\lambda = c \int \frac{d\phi}{R_p^2 \cos^2 \phi} + k. \tag{3.3.10}$$

Substituting Eq. (3.2.9) into Eq. (3.3.10), we have

$$\lambda = c \int (\frac{1 - e^2 \sin^2 \phi}{a^2 \cos^2 \phi}) d\phi + k$$

$$= c \int [\frac{1 - e^2 (1 - \cos^2 \phi)}{a^2 \cos^2 \phi}] d\phi + k$$

$$= c \int (\frac{1 - e^2}{a^2 \cos^2 \phi} + \frac{e^2}{a^2}) d\phi + k$$

$$= c [(\frac{1 - e^2}{a^2}) \tan \phi + \frac{e^2}{a^2} \phi] + k. \tag{3.3.11}$$

Evaluate Eq. (3.3.11) at P_1 and P_2, and substract to eliminate k, thus

$$\lambda_1 = c[(\frac{1 - e^2}{a^2}) \tan \phi_1 + \frac{e^2}{a^2} \phi_1] + k$$

$$\lambda_2 = c [(\frac{1 - e^2}{a^2}) \tan \phi_2 + \frac{e^2}{a^2} \phi_2] + k$$

$$\lambda_2 - \lambda_1 = c [(\frac{1 - e^2}{a^2}) (\tan \phi_2 - \tan \phi_1) + \frac{e^2}{a^2} (\phi_2 - \phi_1)]$$

$$c = \frac{\lambda_2 - \lambda_1}{(\frac{1 - e^2}{a^2})(\tan \phi_2 - \tan \phi_1) + \frac{e^2}{a^2}(\phi_2 - \phi_1)}. \qquad (3.3.12)$$

Substituting Eqs. (3.3.9), (3.2.16) and (3.3.12) into Eq. (3.3.9), we have

$$s = \int_{\phi_1}^{\phi_1} \{\frac{a^2(1 - e^2)}{(1 - e^2\sin^2\phi)^3} + (\frac{1 - e^2\sin^2\phi}{a^2 \cos^2\phi})$$

$$\times \frac{(\lambda_2 - \lambda_1)^2}{[(\frac{1 - e^2}{a^2})(\tan \phi_2 - \tan \phi_1) - \frac{e^2}{a^2}(\phi_2 - \phi_1)]^2}\}^{1/2} d\phi$$

$$s = a \int_{\phi_1}^{\phi_2} \{\frac{(1 - e^2)^2}{(1 - e^2\sin^2\phi)^3}$$

$$+ \frac{(\lambda_2 - \lambda_1)^2(1 - e^2\sin^2\phi)}{[(1 - e^2)(\tan \phi_2 - \tan \phi_1) - e^2(\phi_2 - \phi_1)]^2 \cos^2\phi}\}^{1/2} d\phi.$$

$$(3.3.13)$$

Faced with a non-trivial integral such as Eq. (3.3.13), the only reasonable procedure is a numerical integration on a computer for a specific choice of starting and ending points. The equation itself is completely general.

Several features of the geodesic are now noted. In general, the geodesic is not a plane curve. However, the plane which contains any three near points on it also contains the normal to the spheroid at the center point of the three. The meridianal ellipse is one particular geodesic, and is obtained by setting $d\lambda/d\phi = 0$ in Eq. (3.3.7). The equator is also a particular geodesic. The meridians and the equator are the only geodesics which are plane curves.

Another feature, is that along any geodesic, $R_p \cos \phi \sin \alpha$ is constant, where α is the azimuth, as in Figure 3.3.1.

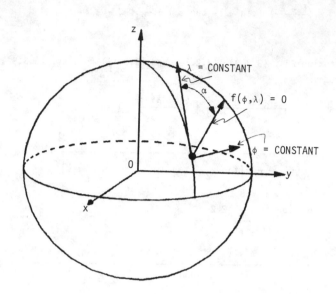

Figure 3.3.1   The Azimuth of a Curve

Three differential formulas will be derived which apply at any point on the geodesic, and relate ϕ, λ, α, and s where α is the azimuth.

$$\frac{d\phi}{ds} = \frac{\cos \alpha}{R_m} \qquad\qquad (3.3.14)$$

$$\frac{d\lambda}{ds} = \frac{1}{R_p} \frac{\sin \alpha}{\cos \phi} \qquad\qquad (3.3.15)$$

$$\frac{d\alpha}{ds} = \frac{1}{R_p} \tan \phi \sin \alpha. \qquad\qquad (3.3.16)$$

Equations (3.3.14) and (3.3.15) can be obtained from a consideration of the angle between a curve on the spheroidal surface, and one of the parametric curves, the meridianal ellipse, where λ is a constant.

From the first fundamental form for the spheroid,

$$E = R_m^2 \tag{3.3.17}$$

$$G = R_p^2 \cos^2 \phi \tag{3.3.18}$$

$$F = 0. \tag{3.3.19}$$

Since $\lambda_1 = c$

$$d\lambda_1 = 0 \tag{3.3.20}$$

and

$$ds_1 = \sqrt{E} \, d\phi_1 \tag{3.3.21}$$

$$\cos \alpha = \frac{E \, d\phi \, d\phi_1 + G \, d\lambda \, d\lambda_1}{ds \, ds_1}. \tag{3.3.22}$$

Substitute Eqs. (3.3.20) and (3.3.21) into Eq. (3.3.22), this gives

$$\cos \alpha = \frac{E}{\sqrt{E}} \frac{d\phi}{ds} \frac{d\phi_1}{d\phi_1}$$

$$= \sqrt{E} \frac{d\phi}{ds}. \tag{3.3.23}$$

Upon substituting Eq. (3.3.17) into Eq. (3.3.23), we have

$$\cos \alpha = R_m \frac{d\phi}{ds}.$$

This establishes Eq. (3.3.14).

$$\sin \alpha = \sqrt{EG} \left(\frac{d\phi_1 \, d\lambda - d\phi \, d\lambda_1}{ds \, ds_1} \right). \tag{3.3.24}$$

Substituting Eqs. (3.3.20) and (3.3.21) into Eq. (3.3.24),

$$\sin \alpha = \frac{\sqrt{EG} \, d\phi_1 \, d\lambda}{\sqrt{E} \, d\phi_1 \, ds}$$

$$= \sqrt{G} \frac{d\lambda}{ds}. \tag{3.3.25}$$

then substituting Eq. (3.3.18) into Eq. (3.3.25), we find

$$\sin \alpha = R_p \cos \phi \frac{d\lambda}{ds}.$$

which establishes Eq. (3.3.15).

From Eqs. (3.3.14) and (3.3.13), the azimuth at P at the initiation of the geodesic can be calculated.

$$\cos \alpha_1 = R_m \frac{d\phi}{ds}$$

$$= (R_m)_1 \left\{ \frac{(1 - e^2)^2}{(1 - e^2 \sin^2 \phi_1)^3} \right.$$

$$\left. + \frac{(\lambda_2 - \lambda_1)^2 (1 - e^2 \sin^2 \phi_1)}{[(1 - e^2)(\tan \phi_2 - \tan \phi_1) - e^2(\phi_2 - \phi_1)]^2 \cos^2 \phi_1} \right\}^{1/2}.$$

(3.3.26)

The rhumbline (or loxodrome) is a curve on the spheroid which meets each consecutive meridian at the same azimuth. From Figure 3.3.2,

$$\tan \alpha = \frac{R_p}{R_n d\phi} \cos \phi \, d\lambda$$

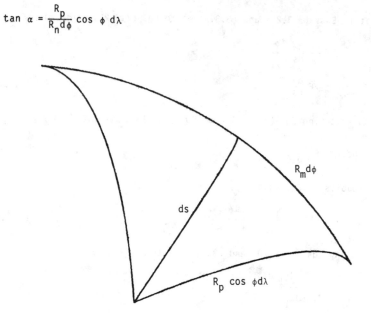

Figure 3.3.2 Differential Element Defining a Rhumbline on a Spheroid

$$d\lambda = \tan \alpha \frac{R_p}{R_m \cos \phi} d\phi.$$

(3.3.27)

Substitute Eqs. (3.2.9) and (3.2.16) into Eq. (3.3.27), to obtain

$$d\lambda = \tan\ \alpha\ \frac{\dfrac{a}{1 - e^2\sin^2\phi}}{\dfrac{a(1 - e^2)\cos\ \phi}{(1 - e^2\sin^2\phi)^{3/2}}}\ d\phi$$

$$= \tan\ \alpha[\frac{1 - e^2\sin^2\phi}{(1 - e^2)\cos\ \phi}]d\phi$$

$$\Delta\lambda = \tan\ \alpha\{\ell n[\tan(\frac{\pi}{4} + \frac{\phi}{2})(\frac{1 - e\ \sin\ \phi}{1 + e\ \sin\ \phi})^{e/2}]\}_{\phi_1}^{\phi_2}. \tag{3.3.28}$$

The kernal of Eq. (3.3.28) arises again when we treat the conformal projections of Chapter 5. The rhumbline, used in conjunction with the Mercator projection (Chapter 5) and the great circle on the gnomonic projection (Chapter 6) are longstanding aids to marine and aerial navigation.

3.4 Geodetic Spheroids [4], [7], [18], [23]

Beginning in the early 1800's, a serious attempt was made to find the correct dimensions of the earth. Geodesists and surveyors undertook to find the best representation.

The physical shape of the earth is too irregular to be used in any mathematical study. Thus, it was necessary to define a fictitious surface which approximates the total shape of the earth. The surfaces of revolution of this section are the geodesists answer to the problem. They are the most convenient surfaces which best fit the true figure of the earth.

The geodesist's first approximation to the shape of the earth is the equipotential surface at mean sea level called the geoid. The geoid is smooth and continuous, and extends under the continents at mean sea level. By definition, the perpendicular at any point of the geoid is in the direction of the gravity vector. This surface, however, is not symmetrical about the axis of revolution, since the distribution of matter within the earth is not uniform. The geoid is the intermediate projection surface between the irregular earth, and the mathematically manageable surface of revolution.

The estimates of the values of the semi-major axis and the flattening have changed over the last 145 years. Progress has meant better instruments, better methods of their use, and better methods of choosing the best fit between the geoid and spheroid. The instruments, their use, and the statistical methods of reducing and fitting the data are beyond the scope of this text. Nevertheless, we must be aware of their existence, and their contribution to mapping.

Table 3.4.1 gives the names and dates of important reference spheroids, as well as the equatorial semi-major axis, a, and the flattening, f, of the ellipse. Historically, the first of these spheroids, the Everest through the Clarke 1880, were intended to fit local areas of the world. Beginning with the Hayford spheroid, an attempt was made to obtain an internationally acceptable representation of the entire world.

Progress has continued in refining the values of the semi-major axis and the flattening. The best representations available today are the World Gravity System of 1972 (WGS-72) and the International Union of Geodesy and Geophysics of 1975 (I.U.G.G.) values. The WGS-72 values will be used in this text.

Unfortunately interest in generating tables such as Table 3.2.1, 3.3.1, 3.3.2, and 4.1.1, and the plotting tables of the projections has flagged since those similar tables incorporating the Clarke 1880 and Hayford spheroids were published [29]. Thus, the tables included in this text, using the WGS-72 spheroid, are the most recent representations of cartographic data available.

Consider now the WGS-72 spheroid. Using Eq. (3.1.4), the eccentricity of the meridianal ellipse is 0.081819. From Eq. (3.1.2)

$$b^2 = a^2(1 - e^2). \tag{3.4.1}$$

Using WGS-72 parameters, b = 6356750 meters. Thus, the difference in length between the equatorial and polar axes is 21385 meters.

Recall from (3.1.4) that the flattening is related to eccentricity and the eccentricity figures in the series expansions developed in this chapter. It has been noted that these expansions are rapidly convergent, due to the small value of e. This is now demonstrated by Table 3.4.2. In the table are powers of e for the WGS-72 spheroid. Note that considering the seven

Table 3.4.1
Reference Spheroids

Reference Spheroid	Date	a (meters)	f
EVEREST	1830	6377304	1/300.8
BESSEL	1841	6377397	1/299.2
AIRY	1858	6377563	1/299.93
CLARKE	1858	6378294	1/294.3
CLARKE	1866	6378206	1/295
CLARKE	1880	6378249	1/293.5
HAYFORD	1910	6378388	1/292.0
KRASOVSKY	1938	6378245	1/298.3
HOUGH	1956	6378270	1/297.0
FISCHER	1960	6378166	1/298.3
KAULA	1961	6378165	1/292.3
I.U.G.G.	1967	6378160	1/298.25
FISCHER	1968	6378150	1/292.3
WGS-72	1972	6378135	1/298.26
I.U.G.G.	1975	6378140	1/298.257

Table 3.4.2
Powers of e for the WGS-72 Spheroid

n	e^n	n	e^n
1	0.081819	4	0.00004481
2	0.00669435	5	0.000003667
3	0.00054772	6	0.000000300

significant figures for a, it is not necessary to carry any expansion beyond e^6.

3.5   Reduction of the Formulas to the Sphere [24]

The most simplified model of the earth is the sphere. In the spherical case, the generatng curve for the surface is the circle, which is an ellipse of zero eccentricity. The semi-major axis and semi-minor axis are the same.

This offers immediate simplifications for the problems of distance, and angular measure. The spheroidal formulas, in general, can be reduced to the spherical by substitution of e = 0. However, the formulas for distance between arbitrary points, and azimuth can best be approached by spherical trigonometry.

The equation for the sphere in Cartesian coordinates is

$$\frac{x^2}{\alpha^2} + \frac{y^2}{a^2} + \frac{z^2}{a^2} = 1.$$ (3.5.1)

Figure 3.5.1 gives the geometry of the spherical earth. Note that the normal to the sphere at a point, P, coincides with the geocentric radius vector. For the sphere, there is only one type of latitude, since geocentric and geodetic latitudes coincide. Longitude is measured in the same way it was for the

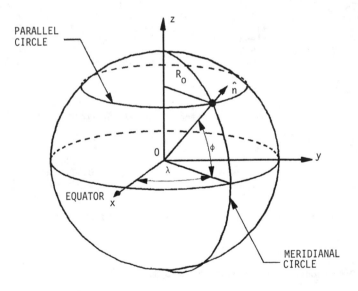

Figure 3.5.1 Geometry of the Sphere

spheroidal case. The sign conventions for latitude and longitude in the spheroidal case also holds for the spherical case.

The radius of a circle of parallel becomes, by substituting e = 0 into Eq. (3.2.7)

$$R_o = a \cos \phi. \tag{3.5.2}$$

By the same substitution, the radii of curvature Eqs. (3.2.9) and (3.2.16) are found to be

$$R_p = R_m = a. \tag{3.5.3}$$

The relation between cartesian and polar coordinates follow from Eq. (3.2.17).

$$\left. \begin{array}{l} x = a \cos \phi \cos \lambda \\ y = a \cos \phi \sin \lambda \\ z = a \sin \phi \end{array} \right\} . \tag{3.5.4}$$

The first fundamental form is, from Eq. (3.2.18)

$$(ds)^2 = a^2(d\phi)^2 + a^2\cos^2\phi(d\lambda)^2. \tag{3.5.5}$$

Distance along the circle of parallel is, from Eq. (3.3.1)

$$d = a\Delta\lambda \cos \phi. \tag{3.5.6}$$

From Eq. (3.3.6), distance along the meridian circle is simply

$$d = a\Delta\phi. \tag{3.5.7}$$

On the sphere, the shortest curve connecting two arbitrary points is an arc of the great circle. The great circle (also called the orthodrome) corresponds to the geodesic curve on the spheroid, but with many simplifications. The great circle is a planar curve which contains the two arbitrary points, and the center of the sphere. The distance on the surface of the sphere, can be obtained from consideration of Figure 3.5.2.

The equations of spherical trigonometry will be used to derive the distance, d. Consider the law of cosines.

$$\cos \theta = \cos (90^0 - \phi_1) \cos (90^0 - \phi_2)$$
$$+ \sin (90^0 - \phi_1) \sin (90^0 - \phi_2) \cos \Delta\lambda$$

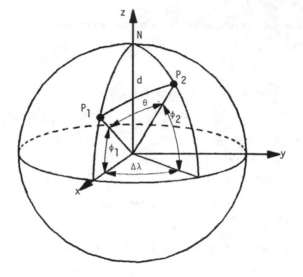

Figure 3.5.2 Distance Between Arbitrary Points on the Sphere

$$= \sin \phi_1 \sin \phi_2$$
$$+ \cos \phi_1 \cos \phi_2 \cos \Delta\lambda. \qquad (3.5.8)$$

Taking the arc-cosine of Eq. (3.5.8)

$$d = a \cos^{-1}[\sin \phi_1 \sin \phi_2 + \cos \phi_1 \cos \phi_2 \cos \Delta\lambda]. \qquad (3.5.9)$$

The azimuth, α, of point P_1 from P_2 is also obtained from the spherical triangle $NP_1 P_2$. Taking the arc-cosine of Eq. (3.5.9) again, the angle θ is now available. Then, the law of sines is applied.

$$\frac{\sin \alpha}{\sin (90^0 - \phi_2)} = \frac{\sin \Delta\lambda}{\sin \theta}$$

$$\sin \alpha = \frac{\cos \phi_2 \sin \Delta\lambda}{\sin \theta}. \tag{3.5.10}$$

Also,

$$\cos \alpha = \cos \theta \cos (90^0 - \phi_1) + \sin \theta \sin (90^0 - \phi_1) \cos (90^0 - \phi_2)$$

$$\cos \alpha = \cos \theta \sin \phi_1 + \sin \theta \cos \phi_1 \sin \phi_2. \tag{3.5.11}$$

From Eqs. (3.5.10) and (3.5.11), the quadrant of the azimuth can be seen. As was mentioned in Chapter 1, azimuth is measured from the North, positive to the East.

The rhumbline or loxodrome is obtained from Eq. (3.3.28) by substituting in e = 0.

$$\Delta\lambda = \tan \alpha [\ln \tan (\frac{\pi}{4} + \frac{\phi_2}{2}) - (\ln \tan (\frac{\pi}{4} + \frac{\phi_1}{2})]. \tag{3.5.12}$$

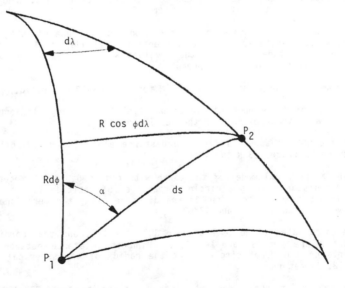

Figure 3.5.3 Differential Element Defining a Rumbline on a Sphere

Equation (3.5.12) can be investigated. If $\phi_1 = \phi_2$, then tan α = 90. This is an azimuth along a parallel circle. If $\lambda_1 = \lambda_2$, tan α = 0, α = 0, yielding a meridian.

The distance along the rhumbline is found from Figure 3.5.3

$$s = \int_{P_1}^{P_2} ds = \int_{P_1}^{P_2} \frac{R}{\cos \alpha} d\phi$$

$$= \frac{R}{\cos \alpha} (\phi_2 - \phi_1). \qquad\qquad (3.5.13)$$

As was mentioned before, the great circle and loxodrome are considered again in conjunction with the Mercator and gnomonic projections.

PROBLEMS

3.1 Calculate the semi-minor axis and eccentricity for the WGS-72 spheroid.

3.2 Calculate the semi-major axis and eccentricity for I.U.G.G. spheroid, and compare this to the values in Problem 3.1.

3.3 Given the I.U.G.G. spheroid. For a geodetic latitude of 45°N, find the geocentric latitude. For a geocentric latitude of 45°N, find the geodetic latitude. Compare these values with corresponding numbers in Table 3.1.1 for the WGS-72 spheroid.

3.4 Find Rm, Rp, and Ro at latitude 37°N for the WGS-72 ellipsoid.

3.5 Find the distance on a circle of parallel for a change in longitude of 10° at a latitude of 43° for the WGS-72 spheroid.

3.6 For the WGS-72 spheroid find the distance along the meridianal ellipse between latitudes 23°N and 47°N.

3.7 Given a spherical model of the earth with radius 6,378,100 meters. What is the distance along a circle of parallel of latitude 43° for a change in longitude of 10°? What is the distance along the meridianal circle between latitudes 23°N and 47°N?

3.8 The coordinates of point 1 are ϕ = 30°, λ = 10°, and the coordinates of point 2 are ϕ = 45° and λ = 70°. Find the distance between the two points on the great circle. Let the radius of the spherical earth be 6,378,100 meters.

3.9 The radius of the spherical earth is taken to be 6,378,100 meters. Let α = 75°. Find the rhumbline distance between latitudes 30°N and 60°N.

4

Equal Area Projections

The requirement which underlies all the projections in this chapter is stated as follows. Every section of the resulting map must bear a constant ratio to the area of the earth represented by it. This requirement is developed below in mathematical terms. Thus, all of the projections of this chapter are founded on some algorithm which maintains the equivalency of area.

As in many endeavors, there is a hard way and an easy way to achieve comparable results. Until recently, cartographers sought to transform directly from the spheroid directly to the developable surface. The easier, and more modern approach, is to transform from the spheroid to the equivalent area sphere, and then transform from the sphere to the developable surface. This results in equations which are far less cumbersome. This chapter follows the modern approach.

First, a transformation is derived which defines an authalic sphere. This authalic sphere has the same total area as the spheroid. The longitude of points is undisturbed by the transformation. However, the transformation requires the definition of an authalic latitude on the sphere, which corresponds to the geodetic latitude on the spheroid. Also, the radius of the authalic sphere must be determined.

Second, positions transformed to the authalic sphere are then transformed onto selected developable surfaces to form a map.

The projections discussed (for the direct transformation from latitude and longitude to cartesian coordinates) are the Conical, Azimuthal, and Cylindrical Equal Area, the Bonne, the Werner, and a selection of world maps: the Sinusoidal, Mollweide, Parabolic, Eumorphic, Eckart, and Hammer-Aitoff. In addition, a simple means to minimize extreme distortion is advanced in the Interrupted projections. Several of these projections are also considered for the inverse transformation from cartesian to geographic coordinates.

A quantitative overview of the theory of distortions is delayed until Chapter 7, where this theory is applied to the most useful of the projections en mass. Plotting tables for selected projections are given.

4.1 Authalic Latitude [2], [8], [22]

Authalic latitude is defined by the equal area projection of the spheroid onto a sphere.

From the fundamental transformation matrix of Eq. (2.7.14), and the condition of equivalency of area:

$$eg = \begin{vmatrix} E' & 0 \\ 0 & G' \end{vmatrix} \begin{vmatrix} \dfrac{\partial \phi_A}{\partial \phi} & \dfrac{\partial \phi_A}{\partial \lambda} \\ \dfrac{\partial \lambda_A}{\partial \phi} & \dfrac{\partial \lambda_A}{\partial \lambda} \end{vmatrix}^2 \tag{4.1.1}$$

In Eq. (4.1.1), ϕ_A and λ_A are the authalic latitude and longitude, respectively, on the authalic sphere, and ϕ and λ are the geodetic latitude and longitude, respectively, on the spheroid. R_A is the radius of the authalic sphere.

On the spheroid, from Eq. (2.3.15), the first fundamental form is:

$$(ds^2) = R_m^2 (d\phi)^2 + R_p^2 \cos^2\phi(d\lambda)^2$$

$$\left. \begin{array}{l} e = R_m^2 \\[2mm] g = R_p^2 \cos^2\phi \end{array} \right\} \quad . \tag{4.1.2}$$

On the authalic sphere, from Eq. (2.3.14), the first fundamental form is:

$$(ds)^2 = R_A^2(d\phi_A)^2 + R_A^2 \cos^2\phi_A(d\lambda_A)^2$$

$$E' = R_A^2 \tag{4.1.3a}$$

$$G' = R_A^2 \cos^2 \phi_A \qquad (4.1.3b)$$

Substitute Eqs. (4.1.2), and (4.1.3), into Eq. (4.1.1).

$$R_m^2 R_p^2 \cos^2 \phi = R_A^4 \cos^2 \phi_A \begin{vmatrix} \dfrac{\partial \phi_A}{\partial \phi} & \dfrac{\partial \phi_A}{\partial \lambda} \\ \dfrac{\partial \lambda_A}{\partial \phi} & \dfrac{\partial \lambda_A}{\partial \lambda} \end{vmatrix}^2 . \qquad (4.1.4)$$

The next step is to evaluate the Jacobian determinant. The longitude is invariant under the transformation: $\lambda = \lambda_A$. Thus,

$$\frac{\partial \lambda_A}{\partial \lambda} = 1 \qquad (4.1.5)$$

$$\frac{\partial \lambda_A}{\partial \phi} = 0. \qquad (4.1.6)$$

The authalic latitude is independent of λ_A.

$$\frac{\partial \phi_A}{\partial \lambda} = 0. \qquad (4.1.7)$$

Substitute Eqs. (4.1.5), (4.1.6), and (4.1.7) into Eq. (4.1.4)

$$R_m^2 R_p^2 \cos^2 \phi = R_A^4 \cos^2 \phi_A \begin{vmatrix} \dfrac{\partial \phi_A}{\partial \phi} & 0 \\ 0 & 1 \end{vmatrix}^2$$

$$R_m R_p \cos \phi = R_A^2 \cos \phi_A \left(\frac{\partial \phi_A}{\partial \phi}\right). \qquad (4.1.8)$$

Equation (4.1.8) can now be separated into an ordinary differential form, (since $\lambda \equiv \lambda_A$).

$$R_m R_p \cos \phi \, d\phi = R_A^2 \cos \phi_A \, d\phi_A. \qquad (4.1.9)$$

Apply the values of R_m and R_p derived in Chapter 2.

$$R_m = \frac{a(1 - e^2)}{(1 - e^2 \sin^2 \phi)^{3/2}} \qquad (2.2.16)$$

$$R_p = \frac{a}{(1 - e^2 \sin^2 \phi)^{1/2}}. \qquad (2.2.17)$$

Substitute Eqs. (2.2.16) and (2.2.17) into Eq. (4.1.9), to obtain

$$\frac{a^2 (1 - e^2)}{(1 - e^2 \sin^2 \phi)^2} \cos \phi \, d\phi = R_A^2 \cos \phi_A \, d\phi_A. \qquad (4.1.10)$$

Integrate Eq. (4.1.10), in which the authalic latitude term is straightforward.

$$\int_0^{\phi_A} R_A^2 \cos \phi_A \, d\phi_A = R_A^2 \sin \phi_A$$

$$= a^2 (1 - e^2) \int_0^{\phi} \frac{\cos \phi}{(1 - e^2 \sin^2 \phi)^2} \, d\phi. \qquad (4.1.11)$$

The easiest way to attack Eq. (4.1.11) is by means of a binomial expansion of the geodetic latitude term.

$$R_A^2 \sin \phi_A = a^2 (1 - e^2) \int_0^{\phi} \cos(1 + 2e^2 \sin^2 \phi + 3e^4 \sin^4 \phi + 4e^6 \sin^6 \phi + \ldots) d\phi$$

$$= a^2 (1 - e^2)(\sin \phi + \frac{2}{3} e^2 \sin^3 \phi + \frac{3}{5} e^4 \sin^5 \phi + \frac{4}{7} e^6 \sin^7 \phi + \ldots)$$

$$(4.1.12)$$

In order to determine R_A, we introduce the condition that $\phi_A = \pi/2$ when $\phi = \pi/2$. Then, Eq. (4.1.12) becomes

$$R_A^2 = a^2 (1 - e^2)(1 + \frac{2}{3} e^2 + \frac{3}{5} e^4 + \frac{4}{7} e^6 + \ldots) \qquad (4.1.13)$$

Equation (4.1.13) gives the radius of the authalic sphere with an area equivalent to that of the spheroid. For example, for the WGS-72 spheroid,

    a = 6,378,135 meters   e = 0.081818

    e^2 = 0.0066941

    e^4 = 0.0000448

    $e^6 \sim 0$

$$R_A = a \sqrt{(1 - e^2)(1 + \frac{2}{3} e^2 + \frac{3}{5} e^4)}$$

$$= (6,378,135) \; \sqrt{(0.9933059)(1 + 0.0044627 + 0.0000268)}$$

$$= 6,371,004 \text{ meters}$$

Substitute Eq. (4.1.13) into Eq. (4.1.12) to obtain the relation between authalic latitude and geodetic latitude.

$$\sin \phi_A = \sin \phi \left[\frac{1 + \frac{2}{3} e^2 \sin^2\phi + \frac{3}{5} e^4 \sin^4\phi + \frac{4}{7} e^6 \sin^6\phi + \cdots}{1 + \frac{2}{3} e^2 + \frac{3}{5} e^4 + \frac{4}{7} e^6 + \cdots} \right] .$$

$$(4.1.14)$$

As was seen in Chapter 3, the eccentricity, e, is a small number for all of the accepted spheroids. Thus, Eq. (4.1.14) contains a rapidly converging series. The relation between authalic and geodetic latitudes is tabulated in Table 4.1.1 for the WGS-72 Reference Ellipsoid, in increments of 5°.

Table 4.1.1
Authalic and Geodetic Latitudes

Geodetic Latitude (Degrees)	Authalic Latitude (Degrees)
0.	0.00000
5.	4.97770
10.	9.95608
15.	14.93577
20.	19.91741
25.	24.90153
30.	29.88863
35.	34.87909
40.	39.87320
45.	44.87114
50.	49.87298
55.	54.87867
60.	59.88802
65.	64.90077
70.	69.91651
75.	74.83477
80.	79.95500
85.	84.97657
90.	90.00000

Radius of the authalic sphere = 6,371,004 meters

Now that the transformation from the spheroid to the sphere is completed, the transformations from the sphere to the developable surfaces is derived in the following sections. In these deviations, the subscript A on the latitude, longitude, and radius of the authalic sphere is dropped, and ϕ, λ, and R are the latitude, longitude, and radius, respectively, of the authalic sphere. The central, or principle meridian is λ_0.

4.2 Conical Projections [2], [22], [24]

Two conical equal area projections are considered. In the first, a cone is tangent to the authalic sphere at a single parallel of latitude. In the second, a cone is secant to a sphere, cutting it at two parallels of latitude. In both cases, the axis of the cone coincides with an extension of the polar axis of the sphere. As is seen in the following paragraphs, the parallels at the points of tangency or secancy are the only true length lines on the map.

Two methods of approach are taken for each of the projections.

The first is the differential geometry approach, which is applied here to the one and two standard parallel cases. The first fundamental form for the authalic sphere is

$$(ds)^2 = R^2(d\phi)^2 + R^2\cos^2\phi(d\lambda)^2$$

$$e = R^2$$
$$g = R^2\cos^2\phi \qquad \cdot \qquad\qquad\qquad (4.2.1)$$

For the polar coordinate system in a plane, the first fundamental form is

$$(ds)^2 = (d\rho)^2 + \rho^2(d\theta)^2$$

$$E' = 1$$
$$G' = \rho^2 \qquad \cdot \qquad\qquad\qquad\qquad (4.2.2)$$

The origin of the projection has coordinates (ϕ_0, λ_0), λ_0 being some longitude on the parallel of tangency. Coordinate λ_0 defines the central meridian of the map, and $\Delta\lambda = \lambda - \lambda_0$.

Impose the conditions that

$$\rho = \rho(\phi) \tag{4.2.3}$$

$$\theta = c_1 \Delta\lambda + c_2. \tag{4.2.4}$$

The constant c_2 is zero, if a further condition is that $\lambda = 0$, when $\theta = 0$.

Imposing the condition of equivalency of area on Eq. (2.7.14)

$$eg = \begin{vmatrix} E' & 0 \\ 0 & G' \end{vmatrix} \begin{vmatrix} \dfrac{\partial\rho}{\partial\phi} & \dfrac{\partial\rho}{\partial\lambda} \\ \dfrac{\partial\theta}{\partial\phi} & \dfrac{\partial\theta}{\partial\lambda} \end{vmatrix}^2. \tag{4.2.5}$$

From Eq. (4.2.3) we have

$$\frac{\partial\rho}{\partial\lambda} = 0. \tag{4.2.6}$$

From Eq. (4.2.4) we obtain

$$\frac{\partial\theta}{\partial\phi} = 0 \tag{4.2.7}$$

$$\frac{\partial\theta}{\partial\lambda} = c_1. \tag{4.2.8}$$

Substitute Eqs. (4.2.1), (4.2.2), (4.2.6), (4.2.7) and (4.2.8) into Eq. (4.2.5).

$$R^4 \cos^2\phi = \rho^2 \begin{vmatrix} \dfrac{\partial\rho}{\partial\phi} & 0 \\ 0 & c_1 \end{vmatrix}^2$$

$$R^2 \cos\phi = -\rho c_1 \left(\frac{\partial\rho}{\partial\phi}\right) \tag{4.2.9}$$

The minus sign is chosen since an increase in ϕ corresponds to a decrease in ρ.

Convert Eq. (4.2.9) into an ordinary differential equation, and integrate; this gives

$$\rho \, d\rho = - \frac{R^2}{c_1} \cos \phi \, d\phi$$

$$\rho^2 = - \frac{2R^2}{c_1} \sin \phi + c_3. \qquad (4.2.10)$$

In Eq. (4.2.10), c_1 will become the constant of the cone, as is shown below, and c_3 will depend on the boundary conditions imposed.

The plane cartesian coordinates of the map are given by:

$$x = \rho \sin \theta$$

$$y = \rho_0 - \rho \cos \theta. \qquad (4.2.11)$$

This development will now be applied to the Conical Equal Area projection with one standard parallel, which is also called Albers' projection. From Chapter 1,

$$\rho_0 = R \cot \phi_0 \qquad (4.2.12)$$

$$\theta = \Delta\lambda \sin \phi_0. \qquad (4.2.13)$$

Comparing Eqs. (4.2.4) and (4.2.13),

$$c_1 = \sin \phi_0. \qquad (4.2.14)$$

The constant c_1 is the constant of the cone (of Chapter 1). Substitute Eq. (4.2.14) into Eq. (4.2.10).

$$\rho^2 = -2R^2 \frac{\sin \phi}{\sin \phi_0} + c_3. \qquad (4.2.15)$$

Evaluate Eq. (4.2.15) at ϕ_0.

$$\rho_0^2 = -2R^2 + c_3$$

$$c_3 = \rho_0^2 + 2R^2. \qquad (4.2.16)$$

Substitute Eq. (4.2.12) into Eq. (4.2.16).

$$c_3 = R^2 \cot^2 \phi_0 + 2R^2$$

$$\quad = R^2(2 + \cot^2 \phi_0). \qquad (4.2.17)$$

Substitute Eq. (4.2.17) into Eq. (4.2.15).

$$\rho^2 = R^2(2 + \cot^2\phi_0 - 2\frac{\sin\phi}{\sin\phi_0})$$

$$= R^2(2\frac{\sin^2\phi_0}{\sin^2\phi_0} + \frac{\cos^2\phi_0}{\sin^2\phi_0} - 2\frac{\sin\phi\sin\phi_0}{\sin^2\phi_0})$$

$$= \frac{R^2}{\sin^2\phi_0}(1 + \sin^2\phi_0 - 2\sin\phi\sin\phi_0)$$

$$\rho = \frac{R}{\sin\phi_0}\sqrt{1 + \sin^2\phi_0 - 2\sin\phi\sin\phi_0}. \qquad (4.2.18)$$

The cartesian plotting equations have been derived. From Eqs. (4.2.11), (4.2.12), and (4.2.13), and including Eq. (4.2.19)

$$x = S[\rho\sin(\Delta\lambda\sin\phi_0)]$$

$$y = S[R\cot\phi_0 - \rho\cos(\lambda\sin\phi_0)] \qquad (4.2.19)$$

where S is the scale factor, and $\Delta\lambda = \lambda - \lambda_0$.

Equations Eqs. (4.2.18) and (4.2.19) are the basis of the plotting Table 4.2.1, for $\phi_0 = 45^0$ and $\lambda_0 = 0°$. No generalized grid is possible; each map is the choice of the user, based on the chosen parallel of tangency. The grid that resulted for this arbitrary choice is Figure 4.2.1.

Two standard parallels may also be selected for a secant cone. This projection is also called the Albers projection. The radii on the map for the standard parallels is:

$$\rho_1 c_1 = R\cos\phi_1 \qquad (4.2.20)$$

$$\rho_2 c_1 = R\cos\phi_2. \qquad (4.2.21)$$

From Eq. (4.2.4) we obtain

$$\theta = c_1\Delta\lambda = \frac{\Delta\lambda R\cos\phi_1}{\rho_1} = \frac{\Delta\lambda R\cos\phi_2}{\rho_2}. \qquad (4.2.22)$$

Substituting Eqs. (4.2.20) and (4.2.21) into Eq. (4.2.10) yields two equations.

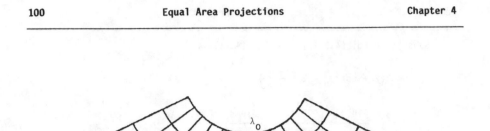

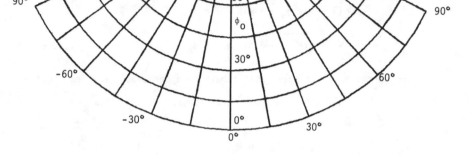

Figure 4.2.1   Conical Equal Area Projection, One Standard Parallel

$$\frac{R^2\cos^2\phi_1}{c_1^2} = -\frac{2R^2}{c_1}\sin\phi_1 + c_3$$

$$R^2\cos^2\phi_1 + 2R^2 c_1\sin\phi_1 - c_1^2 c_3 = 0 \qquad\qquad (4.2.23)$$

$$\frac{R^2\cos^2\phi_2}{c_1^2} = -\frac{2R^2}{c_1}\sin\phi_1 + c_3$$

$$R^2\cos^2\phi_2 + 2R^2 c_1\sin\phi_2 - c_1^2 c_3 = 0. \qquad\qquad (4.2.24)$$

Equations (4.2.23) and (4.2.24) can be solved simultaneously for c_1.

$$R^2\cos^2\phi_1 - R^2\cos^2\phi_2 + 2R^2 c_1\sin\phi_1 - 2R^2 c_1\sin\phi_2 = 0$$

$$\cos^2\phi_1 - \cos^2\phi_2 = 2c_1(\sin\phi_2 - \sin\phi_1)$$

$$c_1 = \frac{\cos^2\phi_1 - \cos^2\phi_2}{2(\sin\phi_2 - \sin\phi_1)}$$

Table 4.2.1
Conical Equal Area Projection, One Standard Parallel

Latitude (Degrees)	Longitude (Degrees)	X (Meters)	Y (Meters)
0.	0.	0.	-4.669
0.	15.	2.033	-4.480
0.	30.	3.997	-3.921
0.	45.	5.825	-3.009
0.	60.	7.453	-1.776
0.	75.	8.827	- .265
0.	90.	9.099	1.473
15.	0.	0.000	-3.227
15.	15.	1.769	-3.063
15.	30.	3.476	-2.576
15.	45.	5.064	-1.784
15.	60.	6.480	- .712
15.	75.	7.675	.602
15.	90.	8.607	2.113
30.	0.	0.000	-1.654
30.	15.	1.479	-1.517
30.	30.	2.906	-1.109
30.	45.	4.235	- .447
30.	60.	5.419	.450
30	75.	6.417	1.548
30.	90.	7.197	2.812
45.	0.	0.000	.000
45.	15.	1.174	.109
45.	30.	2.308	.432
45.	45.	3.363	.959
45.	60.	4.303	1.670
45.	75.	5.096	2.543
45.	90.	5.715	3.546
60.	0.	0.000	1.646
60.	15.	.871	1.727
60.	30.	1.712	1.966
60.	45.	2.495	2.357
60.	60.	3.193	2.885
60.	75.	3.781	3.532
60.	90.	4.240	4.277

Standard parallel: $\phi_0 = 45^0$
Central meridian: $\lambda_0 = 0^0$

$$= \frac{\sin^2 \phi_2 - \sin^2 \phi_1}{2(\sin \phi_2 - \sin \phi_1)}$$

$$= \frac{1}{2} (\sin \phi_2 + \sin \phi_1). \tag{4.2.25}$$

Substituting Eq. (4.2.25) into Eq. (4.2.22), we obtain

$$\theta = \frac{\Delta\lambda}{2} (\sin \phi_1 + \sin \phi_2). \tag{4.2.26}$$

Substitute Eq. (4.2.25) into Eq. (4.2.10)

$$\rho^2 = \frac{4R^2 \sin \phi}{\sin \phi_1 + \sin \phi_2} + c_3. \tag{4.2.27}$$

Evaluate Eq. (4.2.27) at ϕ_1.

$$\rho_1^2 = - \frac{4R^2 \sin \phi_1}{\sin \phi_1 + \sin \phi_2} + c_3$$

$$c_3 = \rho_1^2 + \frac{4R^2 \sin \phi_1}{\sin \phi_1 + \sin \phi_2}. \tag{4.2.28}$$

Substitute Eq. (4.2.28) into Eq. (4.2.27)

$$\rho^2 = \rho_1^2 + 4R^2 \frac{(\sin \phi_1 - \sin \phi)}{\sin \phi_1 + \sin \phi_2} \tag{4.2.29}$$

where, from Eqs. (4.2.20) and (4.2.25),

$$\rho_1 = \frac{2R \cos \phi_1}{\sin \phi_1 + \sin \phi_2}. \tag{4.2.30}$$

A similar development gives

$$\rho^2 = \rho_2^2 + 4R^2 \frac{(\sin \phi_2 - \sin \phi)}{\sin \phi_1 + \sin \phi_2} \tag{4.2.31}$$

$$\rho_2 = \frac{2R \cos \phi_2}{\sin \phi_1 + \sin \phi_2}. \tag{4.2.32}$$

It only remains to substitute into Eqs. (4.2.11) to obtain the plotting equations. One form of these is

$$x = S \cdot \sqrt{\rho_1^2 + 4R^2 \frac{(\sin \phi_1 - \sin \phi)}{\sin \phi_1 + \sin \phi_2}} \sin [\frac{\Delta\lambda}{2} (\sin \phi_1 + \sin \phi_2)] \tag{4.2.33}$$

$$y = S \cdot [\frac{1}{2} (\rho_1 + \rho_2)$$

$$- \sqrt{\rho_1^2 + 4R^2 \frac{(\sin \phi_1 - \sin \phi)}{\sin \phi_1 + \sin \phi_2} \cos [\frac{\Delta\lambda}{2} (\sin \phi_1 + \sin \phi_1)]}\}$$

(4.2.34)

where S is the scale factor and $\Delta\lambda = \lambda - \lambda_0$.

The second form is

$$x = S \cdot \sqrt{\rho_2^2 + \frac{4R^2(\sin \phi_2 - \sin \phi)}{\sin \phi_1 + \sin \phi_2} \sin [\frac{\Delta\lambda}{2} (\sin \phi_1 + \sin \phi_2)]} \quad (4.2.35)$$

$$y = S \cdot \{\frac{1}{2} (\rho_1 + \rho_2)$$

$$- \sqrt{\rho_2^2 + \frac{4R^2(\sin \phi_2 - \sin \phi)}{\sin \phi_1 + \sin \phi_2} \cos [\frac{\Delta\lambda}{2} (\sin \phi_1 + \sin \phi_2)]}\}. \quad (4.2.36)$$

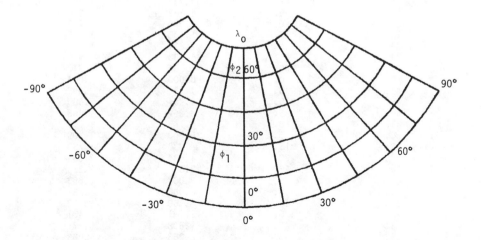

Figure 4.2.2 Conical Equal Area Projection, Two Standard Parallels

The grid is shown in Figure 4.2.2 for the two standard parallel case. The plotting table for the grid is Table 4.2.2. The standard parallels were chosen as $\phi_1 = 30°$, and $\phi_2 = 60°$. The central meridian is $\lambda_0 = 0°$. Again, no completely general grid is possible.

A second approach can be followed for the single standard parallel case. This involves equating corresponding areas on the cone and the sphere.

Consider a cone with constant sin ϕ_0, and let ρ_0 be the radius on the map of the standard parallel ϕ_0. The area on the cone bounded by that parallel is $\pi\rho_0^2$ sin ϕ_0. If ρ is the radius of any other parallel of latitude ϕ, then the area bounded is $\pi\rho^2$ sin ϕ. The area in the strip between these parallels is

Table 4.2.2
Conical Equal Area Projection, Two Standard Parallels

Latitude (Degrees)	Longitude (Degrees)	X (Meters)	Y (Meters)
15.	0.	0.000	-3.324
15.	15.	1.726	-3.169
15.	30.	3.396	-2.710
15.	45.	4.959	-1.961
15.	60.	6.363	- .946
30.	0.	0.000	-1.709
30.	15.	1.438	-1.580
30.	30.	2.831	-1.197
30.	45.	4.133	- .573
30.	60.	5.304	.273
45.	0.	0.000	- .004
45.	15.	1.135	.098
45.	30.	2.234	.400
45.	45.	3.262	.893
45.	60.	4.185	1.560
60.	0.	0.000	1.709
60.	15.	.830	1.783
60.	30.	1.634	2.004
60.	45.	2.386	2.365
60.	60.	3.062	2.853
75.	0.	0.000	3.232
75.	15.	.560	3.282
75.	30.	1.101	3.431
75.	45.	1.608	3.674
75.	60.	2.063	4.003

Standard Parallels $\begin{array}{l}\phi_1 = 30°\\ \phi_2 = 60°\end{array}$

Central Meridian $\lambda_0 = 0°$

$$A = \pi(\rho_0^2 - \rho^2) \sin \phi_0. \qquad (4.2.37)$$

The area on the authalic sphere between the parallels ϕ_0 and ϕ is

$$A = 2\pi R^2(\sin \phi - \sin \phi_0). \qquad (4.2.38)$$

For equal area, equate Eqs. (4.2.37) and (4.2.38)

$$\pi(\rho_0^2 - \rho^2)\sin \phi_0 = 2\pi R^2(\sin \phi - \sin \phi_0)$$

$$(\rho_0^2 - \rho^2)\sin \phi_0 = 2R^2(\sin \phi - \sin \phi_0). \qquad (4.2.39)$$

Substitute Eq. (4.2.12) into Eq. (4.2.39).

$$\sin \phi_0(R^2\cot^2 \phi_0 - \rho^2) = 2R^2(\sin \phi - \sin \phi_0)$$

$$R^2\cot^2 \phi_0 - \rho^2 = 2R^2 \frac{\sin \phi}{\sin \phi_0} - 2R^2$$

$$\rho^2 = R^2\cot^2 \phi_0 + 2R^2 - 2R^2 \frac{\sin \phi}{\sin \phi_0}. \qquad (4.2.40)$$

Equation (4.2.40) is the equivalent of Eq. (4.2.18).

Note the difference in the length of the derivations between the first and second approaches. While the method of differential geometry seems more tedious, it will pay dividends in Chapter 7. The equations for a quantitative estimate of distortion is seen to follow from the differential geometry approach. The second means of derivation leaves the reader without a convenient way of exploring distortions.

Similarly, the case of Albers projection with two standard parallels can be handled with an alternate approach of equating corresponding areas on the cone and the sphere.

Let ϕ_1 and ϕ_2 be the latitude of the two standard parallels, and ρ_1 and ρ_2 be their respective radii on the projection. Let ϕ_2 be greater than ϕ_1. The constant of the cone is c.

The area of the strip of the cone between these latitudes is

$$A = c\pi(\rho_1^2 - \rho_2^2). \qquad (4.2.41)$$

The area of a zone of the authalic sphere between the given latitudes is

$$A = 2\pi R^2 (\sin \phi_2 - \sin \phi_1). \tag{4.2.42}$$

For the equal area projection, equate Eqs. (4.2.41) and (4.2.42).

$$c\pi(\rho_1^2 - \rho_2^2) = 2\pi R^2 (\sin \phi_2 - \sin \phi_1)$$

$$c(\rho_1^2 - \rho_2^2) = 2R^2 (\sin \phi_2 - \sin \phi_1). \tag{4.2.43}$$

Since the standard parallels are true length, we can equate these parallels on the map, and on the authalic sphere

$$2\pi\rho_1 c = 2\pi R\cos \phi_1 \quad , \quad \rho_1 = \frac{R \cos \phi_1}{c} \tag{4.2.44}$$

$$2\pi\rho_2 c = 2\pi R \cos \phi_2 \quad , \quad \rho_2 = \frac{R \cos \phi_2}{c} \tag{4.2.45}$$

Substitute Eqs. (4.2.44) and (4.2.45) into Eq. (4.2.43).

$$c\left(\frac{R^2\cos^2 \phi_1}{c^2} - \frac{R^2\cos^2 \phi_2}{c}\right) = 2R^2(\sin \phi_2 - \sin \phi_1)$$

$$\frac{\cos^2 \phi_1 - \cos^2 \phi_2}{c} = 2(\sin \phi_2 - \sin \phi_1)$$

$$c = \frac{\cos^2 \phi_1 - \cos^2 \phi_2}{2(\sin \phi_2 - \sin \phi_1)}$$

$$= \frac{\sin^2 \phi_2 - \sin^2 \phi_1}{2(\sin \phi_2 - \sin \phi_1)}$$

$$= \frac{1}{2} (\sin \phi_1 + \sin \phi_2).$$

We have now reproduced Eq. (4.2.26).

Substituting the constant of the cone into Eqs. (4.2.44) and (4.2.45)

$$\rho_1 = \frac{2R \cos \phi_1}{\sin \phi_1 + \sin \phi_2}$$

$$\rho_2 = \frac{2R \cos \phi_2}{\sin \phi_1 + \sin \phi_2}.$$

This has reproduced Eqs. (4.2.30) and (4.2.32).

To find the value of ρ for a general latitude ϕ, again, we equate the area on the map and the area on the authalic sphere.

$$c \pi (\rho^2 - \rho_1^2) = 2\pi R^2 (\sin \phi_1 - \sin \phi)$$

$$\rho^2 = \rho_1^2 + 2R^2 (\sin \phi_1 - \sin \phi)$$

$$= \rho_1^2 + \frac{4R^2 (\sin \phi_1 - \sin \phi)}{\sin \phi_1 + \sin \phi_2}.$$

This reproduces Eq. (4.2.29).

The Albers projection has been used extensively in geographic atlases to portray areas of large east to west dimensions. Its best application is in the mid latitudes, and it has been successfully used for maps of the United States. The reason for this is that distortion is a functional latitude, and not longitude. A quantitative estimate of the distortion inherent in this projection is given in Chapter 7.

An example of the inverse transformation from Cartesian to geographic coordinates will be given for the equal area conical projection with one standard parallel. The derivation begins with Eqs. (4.2.19), which are repeated below.

$$X = S[\rho \sin (\Delta\lambda \sin \phi_0)]$$

$$Y = S[R \cot \phi_0 - \rho \cos (\Delta\lambda \sin \phi_0)] \qquad (4.2.19)$$

Rewrite Eqs. (4.2.19) in a form conducive to summing the squares of the trigonometric functions.

$$\frac{x}{S\rho} = \sin (\Delta\lambda \sin \phi_0)$$

$$\frac{SR \cot \phi_0 - y}{S\rho} = \cos (\Delta\lambda \sin \phi_0) \qquad (4.2.46)$$

$$\sin^2(\Delta\lambda \sin \phi_0) + \cos^2(\Delta\lambda \sin \phi_0) = 1$$

$$= (\frac{x}{S\rho})^2 + (\frac{SR \cot \phi_0 - y}{S\rho})^2$$

$$\rho^2 = \frac{1}{S^2} [x^2 + (SR \cot \phi_0 - y)^2] \qquad (4.2.47)$$

$$\rho = \frac{1}{S} \sqrt{x^2 + (SR \cot \phi_0 - y)^2} \qquad (4.2.48)$$

From the first of (4.2.19)

$$\Delta\lambda = \frac{\sin^{-1} (\frac{x}{S\rho})}{\sin \phi_0} \qquad (4.2.49)$$

Equations (4.2.48) and (4.2.49) may be used with $\lambda = \lambda_0 + \Delta\lambda$ to obtain the geographic longitude on the authalic sphere.

Equate the square of Eqs. (4.2.18) and (4.2.47)

$$\frac{1}{S^2} [x^2 + (SR \cot \phi_0 - y)^2]$$

$$= \frac{R^2}{\sin^2 \phi_0} (1 + \sin^2 \phi_0 - 2 \sin \phi \sin \phi_0)$$

$$\frac{\sin^2 \phi_0}{S^2 R^2} [x^2 + (SR \cot \phi_0 - y)^2] = 1 + \sin^2 \phi_0 - 2\sin \phi \sin \phi_0$$

$$\sin \phi = \frac{1}{2\sin \phi_0} \{1 + \sin^2 \phi_0 + \frac{\sin^2 \phi_0}{S^2 R^2} (x^2 + (SR \cot \phi_0 - y)^2]\}$$

$$\phi = \sin^{-1} \{\frac{1}{2\sin \phi_0} + \frac{\sin \phi_0}{2} + \frac{\sin \phi_0}{2S^2 R^2} [x^2 + (SR \cot \phi_0 - y)^2]\} \qquad (4.2.50)$$

Thus, Eq. (4.2.50) gives the authalic latitude on the authalic sphere.

A similar procedure can be applied to the case of two standard parallels. The derivation of this case is included in the problem set.

4.3 Azimuthal Projection [2], [22], [24]

The Azimuthal Equal Area projections may be polar, equatorial, or oblique projections of the authalic sphere directly onto a plane. Two methods of derivation of the polar case will be considered. Then, it will be noted how the oblique and equatorial cases can be obtained by the rotational transformations of Chapter 2.

The Azimuthal Equal Area polar projection (also called the Lambert Azimuthal Equivalent projection) can easily be obtained from the Conical projection with one standard parallel, by setting $\phi_0 = 90°$ in Eqs. (4.2.13) and (4.2.16). Recall that $\Delta\lambda = \lambda - \lambda_0$

$$\theta = \Delta\lambda \qquad (4.3.1)$$

$$\rho = R\sqrt{2(1 - \sin \phi)}. \qquad (4.3.2)$$

The plotting equations, in cartesian coordinates, are, including the scale factor, S,

$$x = R \cdot S \sqrt{2(1 - \sin \phi)}\sin \Delta\lambda \qquad (4.3.3)$$

$$y = -R \cdot S \sqrt{2(1 - \sin \phi)}\cos \Delta\lambda \qquad (4.3.4)$$

The result of plotting these formulas (plotting Table 4.3.1) is Figure 4.3.1. The parallels are concentric circles, unevenly spaced, and the

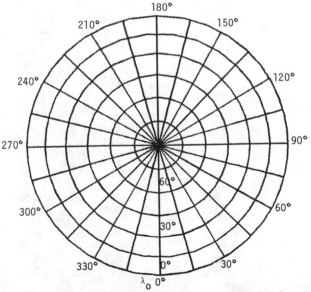

Figure 4.3.1 Azimuthal Equal Area Projection, Polar Case

Table 4.3.1
Azimuthal Equal Area Projection, Polar Case

Latitude Degrees	Longitude Degrees	X Meter	Y Meter
0.	0.	9.020	.000
0.	15.	8.713	2.335
0.	30.	7.812	4.510
0.	45.	6.378	6.378
0.	60.	4.510	7.812
0.	75.	2.334	8.713
0.	90.	.000	9.020
15.	0.	7.766	.000
15.	15.	7.501	2.010
15.	30.	6.725	3.883
15.	45.	5.491	5.491
15.	60.	3.883	5.725
15.	75.	2.010	6.501
15.	90.	.000	7.766
30.	0.	6.378	0.000
30.	15.	6.161	1.651
30.	30.	5.524	3.189
30.	45.	4.510	4.510
30.	60.	3.189	5.524
30.	75.	1.651	6.161
30.	90.	.000	6.378
45.	0.	4.882	0.000
45.	15.	4.715	1.263
45.	30.	4.228	2.441
45.	45.	3.452	3.452
45.	60.	2.441	4.228
45.	45.	1.263	4.715
45.	90.	.000	4.882
60.	0.	3.301	0.000
60.	15.	3.189	.854
60.	30.	2.659	1.651
60.	45.	2.334	2.335
60.	60.	1.651	2.859
60.	75.	.854	3.189
60.	90.	.000	3.301
75.	0.	1.665	0.000
75.	15.	1.604	.431
75.	30.	1.442	.832
75.	45.	1.177	1.177
75.	60.	.832	1.442
75.	75.	.431	1.608
75.	90.	.000	1.865
90.	0.	.000	.000
90.	15.	.000	.000
90.	30.	.000	.000
90.	45.	.000	.000
90.	60.	.000	.000
90.	75.	.000	.000
90.	90.	.000	.000

Point of Tangency: $\phi_0 = 90^0$

Central Meridian: $\lambda_0 = 0^0$

meridians are straight lines. The distortion becomes severe as the equator is reached.

The second way to derive the polar case is equally brief. The area of the segment of the authalic sphere surrounding the pole, and above the latitude ϕ is

$$A = 2\pi R(R - R \sin \phi)$$
$$= 2\pi R^2(1 - \sin \phi). \qquad (4.3.5)$$

This will be transformed into a circle of radius ρ with area

$$A = \pi\rho^2. \qquad (4.3.6)$$

Equating Eqs. (4.3.5) and (4.3.6)

$$\pi\rho^2 = 2\pi R^2(1 - \sin \phi)$$
$$\rho = R\sqrt{2(1 - \sin \phi)}.$$

Equation (4.3.2) has been duplicated.

The oblique and equatorial cases can be obtained from the polar case by applying the rotations of Section 2.10. This has been done in generating the plotting of Tables 4.3.2 and 4.3.3, and Figures 4.3.2 and 4.3.3. For Figure 4.3.3, the point of tangency of the plane against the sphere was arbitrarily chosen at 45° north latitude.

To obtain the oblique variation, write Eqs. (4.3.3) and (4.3.4) in the auxiliary coordinate system.

$$x = R \cdot S \sqrt{2(1 - \sin h)} \sin \alpha \qquad (4.3.7)$$

$$y = R \cdot S \sqrt{2(1 - \sin h)} \cos \alpha. \qquad (4.3.8)$$

From Eqs. (2.10.4) and (2.10.5)

$$\sin h = \sin \phi \sin \phi_p + \cos \phi \cos \phi_p \cos \Delta\lambda \qquad (4.3.9)$$

$$\tan \alpha = \frac{\sin \Delta\lambda}{\cos \phi_p \tan \phi - \sin \phi_p \cos \Delta\lambda}$$

$$\alpha = \tan^{-1}\left(\frac{\sin \Delta\lambda}{\cos \phi_p \tan \phi - \sin \phi_p \cos \Delta\lambda}\right). \qquad (4.3.10)$$

Table 4.3.2
Azimuthal Equal Area Projection, Oblique Case

Latitude	Longitude	X Meters	Y Meter
0.	0.	.000	-4.882
0.	30.	3.552	-4.350
0.	60.	6.714	-2.741
0.	90.	9.020	.000
30.	0.	.000	-1.665
30.	30.	2.846	-1.162
30.	60.	5.251	.332
30.	90.	6.714	2.741
60.	0.	.000	1.665
60.	30.	1.628	1.994
60.	60.	2.920	2.937
60.	90.	3.552	4.350
90.	0.	.000	4.882
90.	30.	.000	4.882
90.	60.	.000	4.882
90.	90.	.000	4.882

Point of Tangency $\phi_0 = 45^0$

Central Meridian $\lambda_0 = 0^0$

Substitute Eqs. (4.3.9) and (4.3.10) into Eqs. (4.3.7) and (4.3.8).

$$
\left.
\begin{aligned}
x &= R \cdot S \sqrt{2(1 - \sin \phi \sin \phi_p - \cos \phi \cos \phi_p \cos \Delta\lambda)} \\
&\quad \cdot \sin \left[\tan^{-1}\left(\frac{\sin \Delta\lambda}{\cos \phi_p \tan \phi - \sin \phi_p \cos \Delta\lambda}\right)\right] \\
y &= R \cdot S \sqrt{2(1 - \sin \phi \sin \phi_p - \cos \phi \cos \phi_p \cos \Delta\lambda)} \\
&\quad \cdot \cos \left[\tan^{-1}\left(\frac{\sin \Delta\lambda}{\cos \phi_p \tan \phi - \sin \phi_p \cos \Delta\lambda}\right)\right]
\end{aligned}
\right\} \quad . \quad (4.3.11)
$$

To obtain the equatorial azimuthal equal area projection, substitute $\phi_p = 0°$ into (4.3.11)

$$
\left.
\begin{aligned}
x &= R \cdot S \cdot \sqrt{2(1 - \cos \phi \cos \Delta\lambda)} \sin \left[\tan^{-1}\left(\frac{\sin \Delta\lambda}{\tan \phi}\right)\right] \\
y &= R \cdot S \cdot \sqrt{2(1 - \cos \phi \cos \Delta\lambda)} \cos \left[\tan^{-1}\left(\frac{\sin \Delta\lambda}{\tan \phi}\right)\right]
\end{aligned}
\right\} \quad . \quad (4.3.12)
$$

The formulas for the distortion in the polar case are given in Chapter 7.

The polar equal area projection is a good means of displaying statistical data when a polar area is a center of interest. The oblique and equatorial projections are useful for the display of statistical data in a region adjacent to the point of tangency of the plane to the authalic sphere.

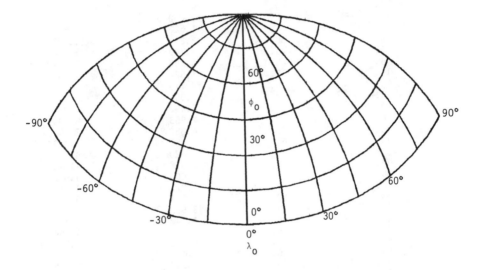

Figure 4.3.2 Azimuthal Equal Area, Oblique Case

As another example of the inverse transformation from Cartesian to geographic coordinates, consider the plotting equations for the azimuthal equal area polar case, Eqs. (4.3.3) and (4.3.4). Divide Eq. (4.3.3) by Eq. (4.3.4)

$$\frac{\sin \Delta\lambda}{\cos \Delta\lambda} = \frac{x}{-y}$$

$$\Delta\lambda = \tan^{-1} \left(\frac{x}{-y}\right) \qquad\qquad (4.3.13)$$

Rearrange (4.3.3)

Table 4.3.3
Azimuthal Equal Area Projection, Equatorial Case

Latitude Degrees	Longitude Degrees	X Meters	Y Meters
0.	0.	.000	.000
0.	15.	1.665	.000
0.	30.	3.302	.000
0.	45.	4.882	.000
0.	60.	6.378	.000
0.	75.	7.766	.000
0.	90.	9.020	.000
15.	0.	0.000	1.665
15.	15.	1.622	1.679
15.	30.	3.215	1.723
15.	45.	4.749	1.800
15.	60.	6.196	1.917
15.	75.	7.527	2.088
15.	90.	8.713	2.335
30.	0.	.000	3.302
30.	15.	1.492	3.328
30.	30.	2.953	3.409
30.	45.	4.350	3.552
30.	60.	5.651	3.768
30.	75.	6.820	4.076
30.	90.	7.812	4.510
45.	0.	.000	4.882
45.	15.	1.272	4.916
45.	30.	2.511	5.023
45.	45.	3.682	5.208
45.	60.	4.748	5.482
45.	75.	5.664	5.864
45.	90.	6.378	6.378
60.	0.	.000	6.378
60.	15.	.959	6.415
60.	30.	1.884	6.526
60.	45.	2.741	6.714
60.	60.	3.493	6.987
60.	75.	4.099	7.351
60.	90.	4.510	7.812
75.	0.	.000	7.766
75.	15.	.540	7.793
75.	30.	1.055	7.875
75.	45.	1.518	8.011
75.	60.	1.902	8.198
75.	75.	2.183	8.435
75.	90.	2.334	8.713
90.	0.	.000	9.020
90.	15.	.000	9.020
90.	30.	.000	9.020
90.	45.	.000	9.020
90.	60.	.000	9.020
90.	75.	.000	9.020
90.	90.	.000	9.020

Point of Tangency: $\phi_0 = 0^0$

Central Meridian: $\lambda_0 = 0^0$

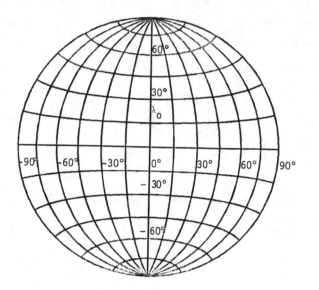

Figure 4.3.3 Azimuthal Equal Area, Equatorial Case

$$\sqrt{2(1 - \sin \phi)} = \frac{x}{RS \sin \Delta\lambda}$$

$$2(1 - \sin \phi) = \frac{x^2}{R^2 S^2 \sin^2 \Delta\lambda}$$

$$\sin \phi = 1 - \frac{x^2}{2R^2 S^2 \sin^2 \Delta\lambda}$$

$$\phi = \sin^{-1}\left(1 - \frac{x^2}{2R^2 S^2 \sin^2 \Delta\lambda}\right) \qquad (4.3.14)$$

Equations (4.3.13) and (4.3.14) with the relation $\lambda = \lambda_0 + \Delta\lambda$ yield authalic longitude and latitude. The inverse transformations for the oblique and equatorial case follow the same procedure, although the mathematics is more messy. These two transformations are left to the problem set.

4.4 Bonne's Projection [2], [8], [22]

Bonne's projection is a modified conical equal area projection. The only straight line in the map is the central meridian, which is a section of the cone and the central meridianal plane.

On the authalic sphere, the first fundamental form is

$$(ds)^2 = R^2(d\phi)^2 + R^2\cos^2\phi(d\lambda)^2$$

$$\left.\begin{array}{l} e = R^2 \\[2em] g = R^2\cos^2\phi \end{array}\right\} . \qquad\qquad (4.4.1)$$

In the plane, a polar coordinate system has the first fundamental form

$$(ds)^2 = (d\rho)^2 + \rho^2(d\theta)^2$$

$$\left.\begin{array}{l} E' = 1 \\[2em] G' = \rho^2 \end{array}\right\} . \qquad\qquad (4.4.2)$$

Applying the condition for equivalency of area the transformation is

$$eg = \begin{vmatrix} E' & 0 \\ 0 & G' \end{vmatrix} \begin{vmatrix} \dfrac{\partial\rho}{\partial\phi} & \dfrac{\partial\rho}{\partial\lambda} \\ \dfrac{\partial\theta}{\partial\phi} & \dfrac{\partial\theta}{\partial\lambda} \end{vmatrix}^2 . \qquad\qquad (4.4.3)$$

The radius of the parallel circle of latitude ϕ is

$$\rho = \rho_0 - \int_{\phi_0}^{\phi} R\,d\phi. \qquad\qquad (4.4.4)$$

From Eq. (4.4.4) we obtain

$$\frac{\partial\rho}{\partial\phi} = - R \qquad\qquad (4.4.5a)$$

$$\frac{\partial \rho}{\partial \lambda} = 0 \tag{4.4.5b}$$

Substituting Eqs. (4.4.1), (4.4.2), and (4.4.5) into Eq. (4.4.3).

$$R^4 \cos^2 \phi = \rho^2 \begin{vmatrix} -R & 0 \\ \frac{\partial \theta}{\partial \rho} & \frac{\partial \theta}{\partial \lambda} \end{vmatrix}^2$$

$$= R^2 \rho^2 \left(\frac{\partial \theta}{\partial \lambda}\right)^2$$

$$R\cos\phi = \rho \left(\frac{\partial \theta}{\partial \lambda}\right). \tag{4.4.6}$$

Convert Eq. (4.4.6) into an ordinary differential equation, and integrate.

$$R\cos \phi \, d\lambda = \rho \, d\theta$$

$$\lambda R \cos \phi = \rho\theta - c$$

$$\theta = \frac{\lambda R \cos \phi}{\rho} + c.$$

The constant c is zero if the condition is imposed that $\lambda = 0$ when $\theta = 0$.

$$\theta = \frac{\lambda R \cos \phi}{\rho}. \tag{4.4.7}$$

In Chapter 1, it was noted that for a conical projection, tangent to a sphere,

$$\rho_0 = R \cot \phi_0. \tag{4.4.8}$$

Carrying out the integration of Eq. (4.4.4)

$$\rho = \rho_0 - R(\phi - \phi_0). \tag{4.4.9}$$

Substitute Eq. (4.4.8) into Eq. (4.4.9).

$$\rho = R \cot \phi_0 - R(\phi - \phi_0). \tag{4.4.10}$$

The cartesian plotting coordinates follow simply, with y-axis as the central meridian, and the origin corresponding to ϕ_0, from Eqs. (4.4.7) and (4.4.10).

$$x = \rho S \sin\left(\frac{\Delta \lambda R \cos \phi}{\rho}\right) \tag{4.4.11}$$

$$y = S[R \cot \phi_0 - \rho \cos(\frac{\Delta\lambda R \cos \phi}{\rho})]$$ (4.4.12)

where S is the scale factor and ϕ and ϕ_0 are in radians. $\Delta\lambda = \lambda - \lambda_0$ has been substituted for λ of the derivation.

The Bonne projection cannot be generalized. It applies to a specific case, with a specified standard parallel, ϕ_0. In order to demonstrate the projection, a plotting table for $\phi_0 = 45$ has been placed in Table 4.4.1.

The Bonne projection has been used as a military map by France.

The grid itself is given in Figure 4.4.1. Note the curvature of the meridians, and the fact that the parallels of latitude are concentric circles.

The inverse transformation from cartesian to geographic coordinates on the authalic sphere follow from Eqs. (4.4.11) and (4.4.12)

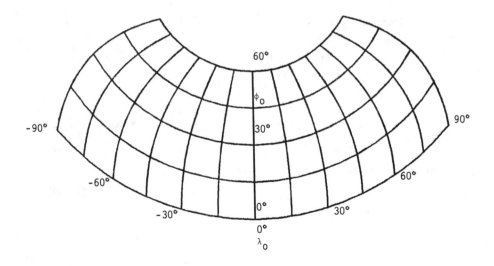

Figure 4.4.1 Bonne's Equal Area Projection

Table 4.4.1
Bonne Equal Area Projection

Latitude Degrees	Longitude Degrees	X Meters	Y Meters
0.	0.	.000	-5.009
0.	15.	1.664	-4.887
0.	30.	3.292	-4.523
0.	45.	4.849	-3.925
0.	60.	6.303	-3.106
0.	75.	7.621	-2.083
0.	90.	8.775	- .879
15.	0.	.000	-3.340
15.	15.	1.606	-3.206
15.	30.	3.167	-2.809
15.	45.	4.641	-2.160
15.	60.	5.988	-1.275
15.	75.	7.170	- .181
15.	90.	8.155	1.094
30.	0.	.000	-1.670
30.	15.	1.438	-1.540
30.	30.	2.830	-1.156
30.	45.	4.131	- .529
30.	60.	5.299	.321
30.	75.	6.296	1.366
30.	90.	7.091	2.572
45.	0.	.000	.000
45.	15.	1.174	.109
45.	30.	2.308	.432
45.	45.	3.363	.959
45.	60.	4.303	1.670
45.	75.	5.096	2.543
45.	90.	5.715	3.546
60.	0.	.000	1.670
60.	15.	.831	1.744
60.	30.	1.635	1.963
60.	45.	2.388	2.320
60.	60.	3.066	2.805
60.	75.	3.649	3.402
60.	90.	4.116	4.093

Standard Parallel: $\phi_0 = 45°$

Central Meridian: $\lambda_0 = 0°$

$$\frac{x}{\rho S} = \sin\left(\frac{\Delta\lambda R \cos\phi}{\rho}\right) \qquad (4.4.13)$$

$$\frac{SR\cot\phi_0 - y}{\rho S} = \cos\left(\frac{\Delta\lambda R \cos\phi}{\rho}\right) \qquad (4.4.14)$$

Square and sum Eqs. (4.4.13) and (4.4.14).

$$\sin^2(\frac{\Delta\lambda R\cos\ \phi}{\rho}) + \cos^2(\frac{\Delta\lambda R\cos\ \phi}{\rho}) = 1$$

$$= (\frac{x}{\rho S})^2 + (\frac{SR\cot\ \phi_0 - y}{\rho S})^2$$

$$\rho^2 = \frac{1}{S^2}[x^2 + (SR\cot\ \phi_0 - y)^2]$$

$$\rho = \frac{1}{S}\ x^2 + (SR\cot\ \phi_0 - y) \qquad\qquad (4.4.15)$$

Rearrange (4.4.13)

$$\Delta\lambda = \frac{\rho}{R\cos\ \phi}\ \sin^{-1}\ (\frac{x}{\rho S}) \qquad\qquad (4.4.16)$$

Equate Eqs. (4.4.13) and (4.4.15), and solve for ϕ.

$$R\cot\ \phi_0 - R(\phi - \phi_0)$$

$$= \frac{1}{S}\ \sqrt{x^2 + (SR\cot\ \phi_0 - y)^2}$$

$$\phi = \phi_0 + \cot\ \phi_0 - \frac{1}{RS}\ \sqrt{x^2 + (SR\cot\ \phi_0 - y)^2} \qquad\qquad (4.4.17)$$

Equations (4.4.15), (4.4.16), and (4.4.17) are used to implement the inverse transformation.

4.5 Cylindrical Projection [2], [15], [22], [24]

In this projection, the meridians and parallels are straight lines, perpendicular to one another. The lines representing the meridians are equally spaced along the equator. It remains to space the parallels to the requirement that an area on the projection is equal to the corresponding area on the authalic sphere.

Since the meridians are perpendicular to the equator, and equally spaced, the abscissa is

$$x = R \cdot S \cdot \Delta\lambda \qquad\qquad (4.5.1)$$

where S is the scale factor, and $\Delta\lambda = \lambda - \lambda_0$, in radians. The longitude of the central meridian is λ_0.

Consider the projection from the authalic sphere to a plane, rectangular coordinate system. For the plane, the first fundamental form is

$$(ds)^2 = (dx)^2 + (dy)^2$$

$$\left. \begin{array}{l} E = 1 \\ G = 1 \end{array} \right\} . \tag{4.5.2}$$

For the sphere, the first fundamental form is

$$(ds)^2 = R^2(d\phi)^2 + R^2\cos^2(d\lambda)^2$$

$$\left. \begin{array}{l} e = R^2 \\ g = R^2\cos^2\phi \end{array} \right\} . \tag{4.5.3}$$

Applying the condition for equivalency of area to the transformation of Chapter 2.

$$eg = \begin{vmatrix} E & 0 \\ 0 & G \end{vmatrix} \begin{vmatrix} \frac{\partial y}{\partial \phi} & \frac{\partial y}{\partial \lambda} \\ \frac{\partial x}{\partial \phi} & \frac{\partial x}{\partial \lambda} \end{vmatrix}^2 \tag{4.5.4}$$

From Eq. (4.5.1), and omitting the scale factor,

$$\frac{\partial x}{\partial \phi} = 0 \tag{4.5.5}$$

$$\frac{\partial x}{\partial \lambda} = R. \tag{4.5.6}$$

Substituting Eqs. (4.5.2), (4.5.3), (4.5.5), and (4.5.6) into Eq. (4.5.4)

$$R^2\cos^2\phi = \begin{vmatrix} \frac{\partial y}{\partial \phi} & \frac{\partial y}{\partial \lambda} \\ 0 & R \end{vmatrix}^2$$

$$= R^2 \left(\frac{\partial y}{\partial \phi}\right)^2$$

$$\frac{\partial y}{\partial \phi} = R \cos \phi. \qquad\qquad\qquad (4.5.7)$$

Integrating Eq. (4.5.7), converted to an ordinary differential equation

$$y = R \sin \phi + c. \qquad\qquad\qquad (4.5.8)$$

The constant c in (4.5.8) can be made to be zero by selecting the origin on the equator. Including the scale factor Eq. (4.5.8) becomes

$$y = R \cdot S \sin \phi. \qquad\qquad\qquad (4.5.9)$$

The same result can be obtained in a different manner. The area of the zone below latitude ϕ on the authalic sphere is

$$A = 2\pi R^2 \sin \phi. \qquad\qquad\qquad (4.5.10)$$

The area on a cylinder tangent to this sphere at the equator is

$$A = 2\pi R y. \qquad\qquad\qquad (4.5.11)$$

Equating Eqs. (4.5.10) and (4.5.11)

$$2\pi R y = 2\pi R^2 \sin \phi$$

$$y = R \sin \phi.$$

This duplicates Eq. (4.5.9). The second method was the one originally employed, and gave the projection its name. The grid resulting from Eqs. (4.5.1) and (4.5.9) is given in Figure 4.5.1. Observe that distortion is intense at higher latitudes, and the projection can be of real service only near the equator. This consideration has limited the usefulness of the projection. Table 4.5.1 gives the plotting coordinates. This projection is also graphically constructed.

This projection can be made oblique or transverse by applying the rotation formulas of Chapter 2. If this is done, the area adjacent to the great circle tangent to the cylinder has a region fairly free of distortion.

The inverse transformation from cartesian to authalic coordinates follows quite easily from Eqs. (4.5.1) and (4.5.9)

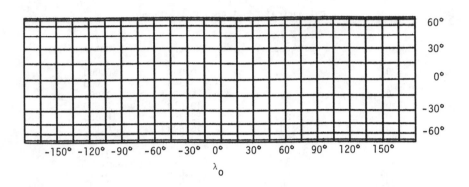

Figure 4.5.1 Cylindrical Equal Area Projection

$$\left.\begin{array}{l} \Delta\lambda = \dfrac{x}{RS} \\[2mm] \phi = \sin^{-1}(\dfrac{y}{RS}) \end{array}\right\}$$ (4.5.12)

4.6 Sinusoidal Projection [8], [15], [20]

The Sinusoidal projection, also called the Sanson-Flamsteed projection, is a projection of the entire authalic sphere. Essentially, it is derived from the Bonne projection by setting $\phi_0 = 0$. Then, ρ approaches infinity. However, we shall use the differential geometry approach.

On the authalic sphere, the first fundamental form is

$$(ds)^2 = R^2(d\phi)^2 + R^2\cos^2\phi(d\lambda)^2$$

$$
\left.\begin{array}{l}
e = R^2 \\
g = R^2\cos^2\phi
\end{array}\right\} \quad .
\qquad (4.6.1)
$$

Table 4.5.1
Cylindrical Equal Area Projection

Latitude Degrees	Longitude Degrees	X Meters	Y Meters
0.	0.	0.000	0.000
0.	30.	3.340	0.000
0.	60.	6.679	0.000
0.	90.	10.019	0.000
0.	120.	13.359	0.000
0.	150.	16.698	0.000
0.	180.	20.038	0.000
30.	0.	0.000	3.189
30.	30.	3.340	3.189
30.	60.	6.679	3.189
30.	90.	10.019	3.189
30.	120.	13.359	3.189
30.	150.	16.698	3.189
30.	180.	20.038	3.189
60.	0.	0.000	5.524
60.	30.	3.340	5.524
60.	60.	6.679	5.524
60.	90.	10.019	5.524
60.	120.	13.359	5.524
60.	150.	16.698	5.524
60.	180.	20.038	5.524
90.	0.	0.000	6.378
90.	30.	3.340	6.378
90.	60.	6.679	6.378
90.	90.	10.019	6.378
90.	120.	13.359	6.378
90.	150.	16.698	6.378
90.	180.	20.038	6.378

Central Meridian: $\lambda_0 = 0°$

On the plane, in Cartesian coordinates, the first fundamental form is

$$(ds)^2 = (dx)^2 + (dy)^2$$

$$
\left.\begin{array}{l}
E' = 1 \\
G' = 1
\end{array}\right\} \quad .
\qquad (4.6.2)
$$

Applying the equal area condition to the transformation, we have

$$eg = \begin{vmatrix} E & 0 \\ 0 & G \end{vmatrix} \begin{vmatrix} \frac{\partial x}{\partial \phi} & \frac{\partial x}{\partial \lambda} \\ \frac{\partial y}{\partial \phi} & \frac{\partial y}{\partial \lambda} \end{vmatrix}^2 . \tag{4.6.3}$$

Along the meridian

$$y = R\phi. \tag{4.6.4}$$

Therefore,

$$\left.\begin{array}{l} \dfrac{\partial y}{\partial \phi} = R \\[2mm] \dfrac{\partial y}{\partial \phi} = 0 \end{array}\right\} . \tag{4.6.5}$$

Substitute Eqs. (4.6.1), (4.6.2), and (4.6.5) into Eq. (4.6.3) to obtain

$$R^4 \cos^2 \phi = \begin{vmatrix} \frac{\partial x}{\partial \phi} & \frac{\partial x}{\partial \lambda} \\ R & 0 \end{vmatrix}^2$$

$$= R^2 \left(\frac{\partial x}{\partial \lambda}\right)^2$$

$$R \cos \phi = \frac{\partial x}{\partial \lambda}. \tag{4.6.6}$$

Convert Eq. (4.6.6) into an ordinary differential equation, and integrate

$$dx = R \cos \phi \, d\lambda$$

$$x = \lambda R \cos \phi + c.$$

Since $x = 0$, when $\lambda = 0$, $c = 0$.

$$x = \Delta \lambda RS \cos \phi. \tag{4.6.7}$$

From Eq. (4.6.4)

$$y = R \cdot S \cdot \phi. \tag{4.6.8}$$

Figure 4.6.1 is a Sinusoidal projection of the earth. All of the parallels are straight lines. The meridians are sinusoidal curves. The central meridian and the equator are straight lines.

The Sinusoidal projection is used for geographical maps. The distortion at extreme latitudes and longitudes is simply ignored. A plotting table is in Table 4.6.1.

The inverse transformation from cartesian to geographical coordinates follow quie simply from Eqs. (4.6.7) and (4.6.8)

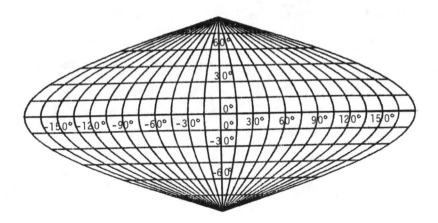

Figure 4.6.1 Sinusoidal Projection

$$\phi = \frac{y}{RS} \tag{4.6.9}$$

$$\Delta\lambda = \frac{x}{RS\cos\phi} \tag{4.6.10}$$

Table 4.6.1
Sinusoidal Projection

Latitude Degrees	Longitude Degrees	X Meters	Y Meters
0.	0.	.000	.000
0.	30.	3.340	.000
0.	60.	6.679	.000
0.	90.	10.019	.000
0	120.	13.359	.000
0	150.	16.698	.000
0	180.	20.038	.000
30.	0.	.000	3.340
30.	30.	2.892	3.340
30.	60.	5.784	3.340
30.	90.	8.677	3.340
30.	120.	11.569	3.340
30.	150.	14.461	3.340
30.	180.	17.353	3.340
60.	0.	.000	6.679
60.	30.	1.670	6.679
60.	60.	3.340	6.679
60.	90.	5.009	6.679
60.	120.	6.679	6.679
60.	150.	8.349	6.679
60.	180.	10.019	6.679
90.	0.	.000	10.019
90.	30.	.000	10.019
90.	60.	.000	10.019
90.	90.	.000	10.019
90.	120.	.000	10.019
90.	150.	.000	10.019
90.	180.	.000	10.019

Central meridian: $\lambda_0 = 0°$

4.7 Mollweide Projection [15], [24]

The Mollweide (Elliptic) projection of the authalic sphere is derived from the construction in Figure 4.7.1. All of the meridians are ellipses. The central meridian is a rectilinear ellipse, or straight line, and the 90° meridians are ellipses of eccentricity zero, or circular arcs. The equator and parallels are straight lines perpendicular to the central meridian. The central meridian and the equator are true length.

The main problem in this projection is spacing the parallels so that the property of equivalence of area is maintained. To do this, apply the law of equal surface from the authalic sphere to the planar map.

The area of the circle centered at 0 is

$$A_1 = \pi r^2.$$ (4.7.1)

This is to be equal in area to a hemisphere

$$A_1 = 2\pi R^2.$$ (4.7.2)

Equating Eqs. (2.7.1) and (4.7.2)

$$\pi r^2 = 2\pi R^2$$

$$r^2 = 2R^2$$ (4.7.3)

$$r = \sqrt{2}\ R.$$ (4.7.4)

Consider Figure 4.7.1. The area between latitude ϕ on the sphere, and the equator is

$$A = 2\pi R^2 \sin\ \phi.$$ (4.7.5)

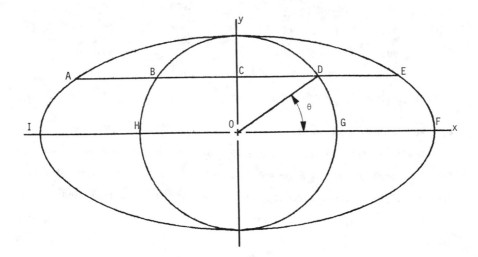

Figure 4.7.1 Geometry of the Molliweide Projection

This area is equal to the area AEFI on the figure. For a circle inscribed within an ellipse, where the radius of the circle is one half of the semi-major axis the area BDGH equals one half of the area AEFI [17]. Consider half of the area BDGH, that is area CDGO. This area is composed of the triangle OCD and the sector ODG. The area of the triangle is

$$A_A = \frac{1}{2} \, r \, \sin \theta \, r \, \cos \theta$$

$$= \frac{r^2}{4} \sin 2\theta. \qquad (4.7.6)$$

The area of the sector is

$$A_S = \frac{r^2 \theta}{2}. \qquad (4.7.7)$$

Equate the spherical area and the map area through Eqs. (4.7.5), (4.7.6), and (4.7.7).

$$2\pi R^2 \sin \phi = 4(\frac{1}{2} \, r^2 \theta + \frac{1}{4} \, r^2 \sin 2\theta)$$

$$\pi R^2 \sin \phi = r^2 \theta + \frac{1}{2} \, r^2 \sin 2\theta. \qquad (4.7.8)$$

Substitute Eq. (4.7.3) into Eq. (4.7.8)

$$\pi R^2 \sin \phi = 2R^2 \theta + R^2 \sin 2\theta$$

$$\pi \sin \phi = 2\theta + \sin 2\theta. \qquad (4.7.9)$$

We are now faced with a transcendental equation to be solved for θ. For limited accuracy, a graph of θ versus ϕ can be constructed, as in Figure 4.7.2, and values of θ read for given values of ϕ. However, for computer implementation of this projection, it is necessary to resort to a numerical solution.

Apply the Newton-Raphson method [14]. Write Eq. (4.7.9) as

$$f(\theta) = \pi \sin \phi - 2\theta - \sin 2\theta = 0. \qquad (4.7.10)$$

Differentiating Eq. (4.7.10)

$$f'(\theta) = -2 - 2 \cos 2\theta. \qquad (4.7.11)$$

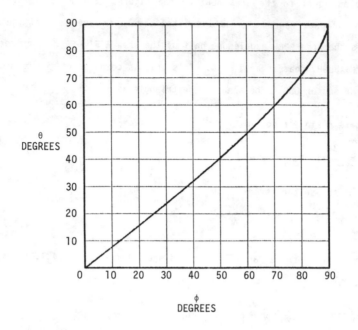

Figure 4.7.2 θ vs. ϕ for the Mollweide Projection

The iterative solution of Eq. (4.7.9) for θ as a function of ϕ is

$$\theta_{n+1} = \theta_n - \frac{f(\theta_n)}{f'(\theta_n)}. \tag{4.7.12}$$

Substitute Eqs. (4.7.10) and (4.7.11) into Eq. (4.7.12).

$$\theta_{n+1} = \theta_n + \frac{\pi \sin \phi - 2\theta_n - \sin 2\theta_n}{2 + 2 \cos 2\theta_n}$$

where θ_n is in radians. This has a rapid convergence if the initial guess for θ is the given value of ϕ.

Once θ is found, the mapping equations quickly follow from Figure 4.7.1.

$$y = r \cdot S \sin \theta$$

$$x = \frac{\Delta\lambda}{180} S \cdot 2r \cos \theta$$

$$= \frac{\Delta\lambda}{90} rS \cos \theta$$

(4.7.13)

S is the scale factor, and $\Delta\lambda = \lambda - \lambda$ is in degrees. The central meridian has longitude λ_0. Substituting Eq. (4.7.4) into Eqs. (4.7.13), we arrive at the plotting equations

$$y = \sqrt{2} R \cdot S \sin \theta$$

$$x = \frac{\Delta\lambda\sqrt{2}}{90} R \cdot S \cdot \cos \theta$$

(4.7.14)

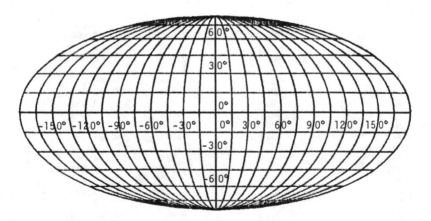

Figure 4.7.3 Mollweide Projection

The result of applying Eq. (4.7.14), and the iteration is the grid of Figure 4.7.3. The distortion towards the poles is not as great as in the Sinusoidal projection, but it is more noticeable than in the Hammer-Aitoff projection. The chief use of the Mollweide projection is for geographical illustrations relating to area, where distortions are not disturbing. This projection gives a reliable representation of statistical data.

Plotting tables for the Mollweide projection are given in Table 4.7.1. Latitudes 0° to 90°, in steps of 30°, and longitudes 0° to 180° in steps of 30°, are tabulated.

The inverse transformation from cartesian coordinates to geographical coordinates on the authalic sphere follows from Eqs. (4.7.9) and (4.7.14)

$$\theta = \sin^{-1} \left(\frac{y}{\sqrt{2}RS}\right) \qquad\qquad (4.7.15)$$

Table 4.7.1
Mollweide Projection

Latitude Degree	Longitude Degree	X Meters	Y Meters
0.	0.	.000	0.000
0.	30.	3.340	0.000
0.	60.	6.679	0.000
0.	90.	10.019	0.000
0.	120.	13.359	0.000
0.	150.	16.698	0.000
0.	180.	20.039	0.000
30.	0.	.000	4.047
30.	30.	3.055	4.047
30.	60.	6.110	4.047
30.	90.	9.165	4.047
30.	120.	12.220	4.047
30.	150.	15.275	4.047
30.	180.	18.330	4.047
60.	0.	.000	7.638
60.	30.	2.161	7.638
60.	60.	4.322	7.638
60.	90.	6.483	7.638
60.	120.	8.644	7.638
60.	150.	10.806	7.638
60.	180.	12.967	7.638
90.	0.	.000	10.018
90.	30.	.000	10.018
90.	60.	.000	10.018
90.	90.	.000	10.018
90.	120.	.000	10.018
90.	150.	.000	10.018
90.	180.	.000	10.018

Central meridian: $\lambda_0 = 0°$

$$\phi = \sin^{-1}(\frac{2\theta + \sin 2\theta}{\pi})$$ (4.7.16)

$$\Delta\lambda = \frac{90x}{\sqrt{2} \text{ RS } \cos \theta}$$ (4.7.17)

4.8 Parabolic Projection [8], [15], [24]

The Parabolic (also called Craster) projection of the entire world is shown in Figure 4.8.1.   The parallels are straight lines parallel to a straight line equator. The meridians are parabolic arcs.

The Parabolic projection can be constructed from consideration of Figure 4.8.2. Let the equator be four arbitrary units of length. Then, the central meridian is two units.   A scale factor will be applied at the end of the derivation to convert the assumed units to the equatorial circumference. R is the radius of the authalic sphere.

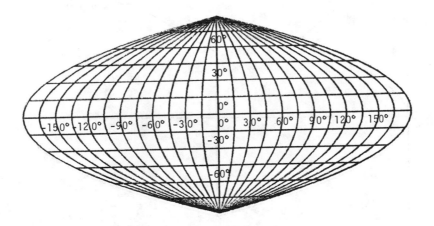

Figure 4.8.1 Parabolic Projection

Consider the cross hatched area in Figure 4.8.2, bounded by an outer meridian, the central meridian, and the equator. The outer meridian is taken to be parabola, $y^2 = x/2$, with its vertex at (0,0). The mapping criterion requires that one quarter of the area on the authalic sphere will be equivalent to the shaded area between $x = 0$, and $x = 2$. Thus, one half of the zone between the equator, and a given parallel, ϕ, will be

$$A = \int_0^y (2 - x)dy$$

$$= \int_0^y (2 - 2y^2)dy$$

$$= 2y - \frac{2}{3} y^3 \Big|_0^y$$

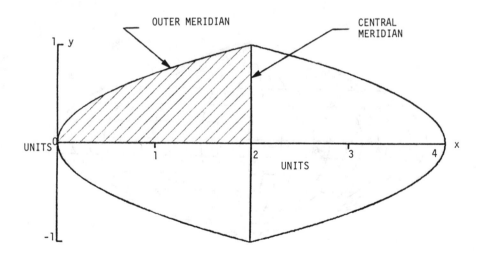

Figure 4.8.2 Geometry for the Parabolic Projection

$$= 2y - \frac{2}{3} y^3. \qquad (4.8.1)$$

The total area of the sphere is $4\pi R^2$. Substituting this value, and $y = 1$ into Eq. (4.8.1)

$$\pi R^2 = 4/3 \qquad (4.8.2)$$

$$R = \frac{4}{3\pi}$$

$$= 0.651470$$

This gives the relation between R and the arbitrary unit on the map. Thus, one map unit is equal to $1.53499R$.

Next, relate the map coordinate to authalic latitude. The area of a zone on the authalic sphere from the equator to latitude ϕ is $2\pi R^2 \sin \phi$. Half of this zone is then

$$A = \pi R^2 \sin \phi. \qquad (4.8.3)$$

Equate Eqs. (4.8.1) and (4.8.3)

$$2y - \frac{2}{3} y^3 = \pi R^2 \sin \phi. \qquad (4.8.4)$$

Substitute Eq. (4.8.2) into Eq. (4.8.4).

$$2y - \frac{2}{3} y^3 = \frac{4}{3} \sin \phi$$

$$y^3 - 3y + 2 \sin \phi = 0.$$

A solution of this transcendental equation is

$$y = 2 \sin \phi/3 \qquad (4.8.5)$$

which can be verified by substitution. A scale factor, S, and radius, R, may be introduced into Eq. (4.8.5) to obtain the ordinate, recalling also that one map unit equals $1.53499R$.

$$y = 3.06998SR \sin \phi/3. \qquad (4.8.6)$$

The abscissa may be obtained by the following development. The length of a parallel between the central meridian and the outer meridian is given by

$$\ell = 2 - 2y^2. \tag{4.8.7}$$

Substitute Eq. (4.8.5) into Eq. (4.8.7) to obtain the length as a function of latitude.

$$\ell = 2(1 - 4 \sin^2 \phi/3)$$

$$= 2(1 + 2 \cos \frac{2\phi}{3} - 2)$$

$$= 2(2 \cos \frac{2\phi}{3} - 1). \tag{4.8.8}$$

The parallels are divided proportionally for the intersections of the meridians. From (4.8.8), and including the scale factor, S, the radius R, and the relation between the map unit and R.

$$x = 1.53499 \frac{\Delta\lambda}{180} \ell S \cdot R$$

$$= 1.53499 \frac{\Delta\lambda}{90} SR(2 \cos \frac{2\phi}{3} - 1). \tag{4.8.9}$$

In Eq. (4.8.9) $\Delta\lambda = \lambda - \lambda_0$ is the difference in longitude between the given meridian and the central meridian, in degrees.

Since this is an equal area projection, its use is for statistical representation. No attempt is made to avoid distortion in angles and shapes. However, the distortion is less than in the Mollweide projection because the meridians and parallels do not intersect at such acute angles. Also, the symmetry and parabolic curves lend a certain aesthetic quality.

Equations (4.8.6) and (4.8.9) have computed in Table 4.8.1. This table gives the longitude from 0° to 180° in steps of 30°, and latitude from 0° to 90° in steps of 30°.

The inverse transformation for the parabolic projection follow from Eqs. (4.8.6) and (4.8.9)

$$\left. \begin{array}{l} \phi = 3 \sin^{-1} (\frac{y}{3.06998RS}) \\[2ex] \Delta\lambda = \dfrac{90x}{1.53499RS(2 \cos \frac{2\phi}{3} - 1)} \end{array} \right\} \tag{4.8.10}$$

Table 4.8.1
Parabolic Projection

Latitude Degrees	Longitude Degrees	X Meters	y Meters
0.	0.	.000	.000
0.	30.	3.340	.000
0.	60.	6.679	.000
0.	90.	10.019	.000
0.	120.	13.359	.000
0.	150.	16.698	.000
0.	180.	20.038	.000
30.	0.	.000	3.480
30.	30.	2.937	3.480
30.	60.	5.874	3.480
30.	90.	8.810	3.480
30.	120.	11.747	3.480
30.	150.	14.684	3.480
30.	180.	17.621	3.480
60.	0.	.000	6.853
60.	30.	1.777	6.853
60.	60.	3.554	6.853
60.	90.	5.331	6.853
60.	120.	7.108	6.853
60.	150.	8.885	6.853
60.	180.	10.662	6.853
90.	0.	.000	10.019
90.	30.	.000	10.019
90.	60.	.000	10.019
90.	90.	.000	10.019
90.	120.	.000	10.019
90.	150.	.000	10.019
90.	180.	.000	10.019

Central meridian: $\lambda_o = 0^0$

4.9 Hammer-Aitoff Projection

The Hammer-Aitoff projection, shown in Figure 4.9.1, is derived by a mathematical manipulation of the Azimuthal Equal Area projection of Section 4.3. In the Hammer-Aitoff projection, the sphere is represented within an ellipse, with semi-major axis twice the length of the semi-minor axis. In this respect, it is similar to the Mollweide projection. However, in the Hammer-Aitoff projection, the parallels are curved lines, rather than straight.

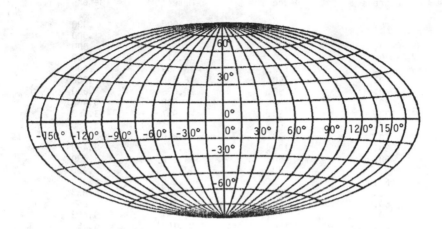

Figure 4.9.1 Hammer-Aitoff Projection

The grid of meridians and parallels is obtained by the orthogonal projection of the Azimuthal Equal Area projection, equatorial case, onto planes making angles of 60° to the plane of the Azimuthal projection.

Figure 4.9.2 demonstrates the means of projection. In this figure, we are looking upon the edges of the planes, which appear as straight lines. Since the angle between the planes is 60°, DO = 2AO, and OB = 2OC. Thus, the total length DO plus OC is the entire equator, as AB is half of the equator. It is assumed that for the Hammer-Aitoff projection the total map of the authalic sphere is obtained by unfolding DO and OC into a plane DOC, with O as the position of the central meridian. In this projection, the ordinate is not modified from a comparable point on the Azimuthal Equal area projection.

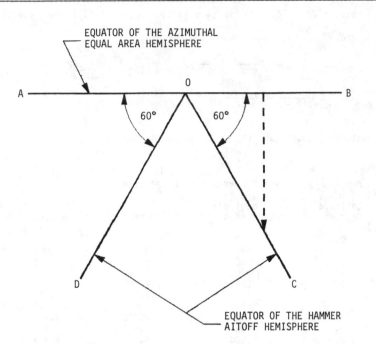

Figure 4.9.2 Geometry of the Hammer-Aitoff Projection

Converting to the coordinates of the auxiliary system Eqs. (4.3.3) and (4.3.4) become

$$x = R \cdot S \cdot \sqrt{2(1 - \sin h)} \, \sin \alpha \qquad (4.9.1)$$

$$y = R \cdot S \cdot \sqrt{2(1 - \sin h)} \, \cos \alpha. \qquad (4.9.2)$$

From Eqs. (2.10.3) and (2.10.5), for $\phi_0 = 0°$

$$\sin h = \cos \phi \cos \lambda_a \qquad (4.9.3)$$

$$\tan \alpha = \frac{\sin \lambda_a}{\tan \phi}$$

$$\alpha = \tan^{-1} \left(\frac{\sin \lambda_a}{\tan \phi} \right) \qquad (4.9.4)$$

where λ_a is the latitude on the azimuthal projection.

Substitute Eqs. (4.9.3) and (4.9.4) into Eqs. (4.9.1) and (4.9.2), and let $\lambda_a = \lambda/2$.

$$\left. \begin{aligned} x &= 2 \cdot R \cdot S \sqrt{2(1 - \cos\phi\,\cos\lambda/2)}\ \sin\left[\tan^{-1}\left(\frac{\sin\lambda/2}{\tan\phi}\right)\right] \\[2ex] y &= R \cdot S \sqrt{2(1 - \cos\phi\,\cos\lambda/2)}\ \cos\left[\tan^{-1}\left(\frac{\sin\lambda/2}{\tan\phi}\right)\right]. \end{aligned} \right\} \quad (4.9.5)$$

Equations (4.9.5) gave the plotting coordinates of Table 4.9.1.

Table 4.9.1
Hammer-Aitoff Projection

Latitude Degrees	Longitude Degrees	X Meters	Y Meters
0.	0.	.000	.000
0.	30.	3.330	.000
0.	60.	6.603	.000
0.	90.	9.763	.000
0.	120.	12.750	.000
0.	150.	15.531	.000
0.	180.	18.040	.000
30.	0.	.000	3.302
30.	30.	2.984	3.328
30.	60.	5.905	3.409
30.	90.	8.700	3.552
30.	120.	11.303	3.768
30.	150.	13.640	4.076
30.	180.	15.623	4.510
60.	0.	.000	6.378
60.	30.	1.917	6.415
60.	60.	3.767	6.526
60.	90.	5.482	6.714
60.	120.	6.987	6.987
60.	150.	8.198	7.351
60.	180.	9.020	7.812
90.	0.	.000	9.020
90.	30.	.000	9.020
90.	60.	.000	9.020
90.	90.	.000	9.020
90.	120.	.000	9.020
90.	150.	.000	9.020
90.	180.	.000	9.020

Central meridian: $\lambda_0 = 0°$

That the area enclosed by the ellipse of the Hammer-Aitoff projection, corresponding to the entire sphere, is twice the area of the Azimuthal projection, corresponding to a hemisphere, follows easily from the geometry

of the ellipse with a circle of the radius of the semi-minor axis inscribed
[17].

In Figure 4.9.1, the central meridian and the equator are the only
straight lines in the grid. The rest of the meridians and parallels are
curves. The curvature of the parallels with respect to the meridians is such
that there is less angular distortion than appears at higher latitudes, and
more distant longitudes in the Mollweide projection.

The Hammer-Aitoff projection is used primarily for statistical
representation of data. The distortion that occurs is overlooked.

4.10 Eckert's Projection [24], [26]

Eckert produced a total of six projections. The one that has received
some fame is Eckert 4. In this projection, the central meridian is half the
length of the equator. From Figure 4.10.1, the equator, parallels, and
central meridian are straight lines. The method of choosing the parallels
requires that the other meridians be elliptical curves.

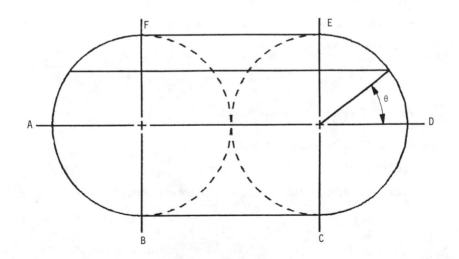

Figure 4.10.1 Geometry for Eckert's Projection

In the projection, the spacing of the parallels decreases with latitude in a manner that makes this an equal area projection. The derivation is similar to that of the Mollweide.

The area of the hemisphere is

$$A_1 = 2\pi R^2. \tag{4.10.1}$$

The area north of the equator on the Eckert projection is, from the Figure 4.13.1,

$$A_1 = 2r^2 + \frac{\pi r^2}{2}. \tag{4.10.2}$$

Equating Eq. (4.10.1) and Eq. (4.10.2)

$$2\pi R^2 = 2r^2 + \frac{\pi r^2}{2}$$

$$= r^2(2 + \frac{\pi}{2}). \tag{4.10.3}$$

Thus, r = 1.32650R

The area on the hemisphere below latitude ϕ is

$$A_2 = 2\pi R^2 \sin \phi. \tag{4.10.4}$$

The corresponding area on the projection, including a rectangle, two sectors, and two triangles, is

$$A_2 = 2r \cdot r \sin \theta + \frac{2}{2} r \sin \theta \cdot r \cos \theta + \frac{2}{2} r^2 \theta$$

$$= r^2(2 \sin \theta + \sin \theta \cos \theta + \theta). \tag{4.10.5}$$

Equating Eq. (4.10.4) and Eq. (4.10.5)

$$2\pi R^2 \sin \phi = r^2(2 \sin \theta + \sin \theta \cos \theta + \theta). \tag{4.10.6}$$

Equate Eq. (4.13.3) and Eq. (4.13.6) to obtain a relation between ϕ and θ.

$$r^2(2 + \frac{\pi}{2}) \sin \phi = r^2(2 \sin \theta + \sin \theta \cos \theta + \theta)$$

$$(2 + \frac{\pi}{2}) \sin \phi = 2 \sin \theta + \sin \theta \cos \theta + \theta. \tag{4.10.7}$$

Again, we have a transcendental equation to be solved for θ as a function of φ. And, again, the Newton-Raphson method will produce the required result [14]. Armed with this, the cartesian plotting coordinate are, recalling the relation between r and R

$$x = 1.32650S \cdot R(1 + \cos\theta)(\Delta\lambda)$$

$$y = 1.32650S \cdot R \cdot \frac{\pi}{2}\sin\theta \qquad\qquad (4.10.8)$$

As with the Mollweide projection, the relation between φ and θ may be plotted. This is given as Figure 4.10.2 and Figure 4.10.3 is the grid produced from Eqs. 4.10.8

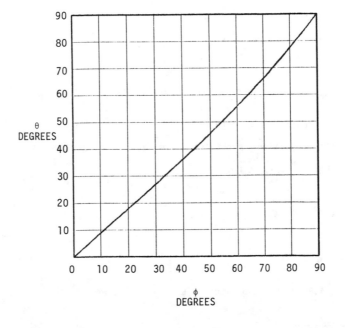

Figure 4.10.2   θ vs φ for Eckert's Projection

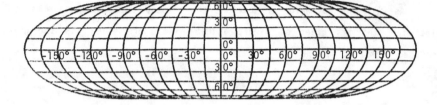

Figure 4.10.3   Eckert's Projection

4.11 Werner's Projection [2], [22]

Werner's projection is obtained from Bonne's projection by letting $\phi_0 = 90°$. From Eq. (4.4.5), we have

$$\rho = - \int_{\pi/2}^{\phi} R \, d\phi$$

$$= R(\pi/2 - \phi). \tag{4.11.1}$$

From Eq. (4.4.8), after substituting Eq. (4.11.1), we obtain

$$\theta = \frac{\Delta\lambda R \cos \phi}{R(\pi/2 - \phi)}$$

$$= \frac{\Delta\lambda \cos \phi}{\pi/2 - \phi}. \tag{4.11.2}$$

If the origin is chosen at the pole, the cartesian plotting equations become

$$x = RS(\pi/2 - \phi)\sin\left(\frac{\Delta\lambda \cos \phi}{\pi/2 - \phi}\right)$$
$$y = -RS(\pi/2 - \phi)\cos\left(\frac{\Delta\lambda \cos \phi}{\pi/2 - \phi}\right) \Bigg\} \qquad (4.11.3)$$

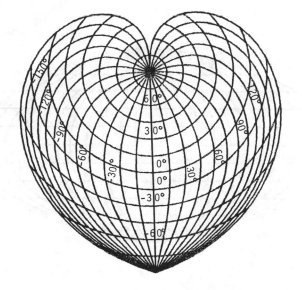

Figure 4.11.1   Werner's Projection

The grid corresponding to Eqs. (4.11.3) is in Figure 4.11.1.   The only straight line in this cardioid shaped projection is the central meridian. Note that distortion becomes excessive at the south pole, and at increased longitude from the central meridian.   The parallels are still concentric circles. S is the scale factor.

4.12.1 Eumorphic Projection [24]

The Eumorphic projection in Figure 4.12.1 is essentially an arithmetic mean between the sinusoidal and Mollweide projectives.   The projection is obtained for each longitude by summing the distance along the central meridian to a given parallel for the Sinusoidal and Mollweide, and then dividing by 2. The length of the parallel is then needed. This is obtained by requiring that the area between the equator, the central meridian, the given meridian, and the parallel under consideration be the same on the Eumorphic as on the

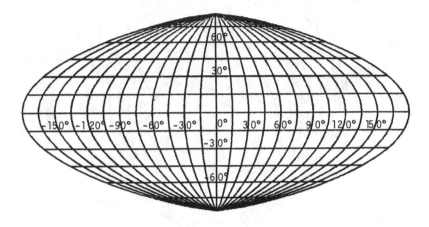

Figure 4.12.1 Eumorphic Projection

Mollweide or the Sinusoidal projection. This means obtaining comparable areas on the Mollweide or the Sinusoidal by either a planimeter or integrating a polynomial which approximates the meridian curve. It is then necessary to obtain the x-coordinates on the Eumorphic by trial and error with a planimeter, or fitting a polynomial curve. This is time consuming, and has not been done in this text. However, Table 4.12.1 gives the y coordinates for the Eumorphic projection.

Table 4.12.1
Parallel Spacings on the Eumorphic Projection

ϕ (Degrees)	Y (Meters)
0	0.000
30	0.547
60	1.061
90	1.490

4.13 Interrupted Projections [22]

Interrupted projections of the authalic sphere are a means of reducing maximum distortion at the expense of continuity of the map. Figures 4.13.1, 4.13.2, 4.13.3, 4.13.4, and 4.13.5 show interrupted Sinusoidal Mollweide, Parabolic, Eumorphoric and Eckert projections, respectively.

Certain meridians are chosen as reference meridians, which are straight lines. The equator is also a straight line. Then other meridians are chosen where the breaks occur. Note that it is not necessary for a reference meridian to appear in both hemispheres. One can choose a half meridian in either hemisphere.

The parallels are spaced in the same manner as in the regular Sinusoidal, Mollweide, Parabolic, Eumorphic, or Eckert projections. The difference comes in the method of handling the spacing of the meridians. Each reference meridian becomes the local axis of the coordinate system, and the abscissa is marked, east or west, until a break is reached. Then, one goes to the next reference meridian, and repeats the process.

This procedure has been applied to the Sinusoidal, Mollweide, Parabolic, Eumorphic and Eckert projections. In each case, the respective plotting equations of the respective sections have been used.

The grids which result from this method are rather exotic in appearance. Distortion, since it is greatest at the farthest longitude from the central meridian, is always significantly decreased. These maps are generally used as statistical representations, so the breaks cannot create

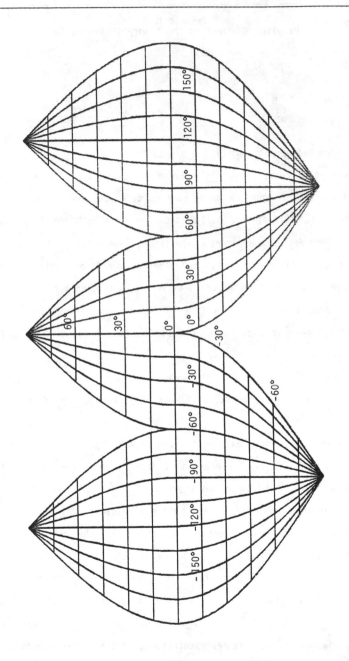

Figure 4.13.1 Interrupted Sinusoidal Projection

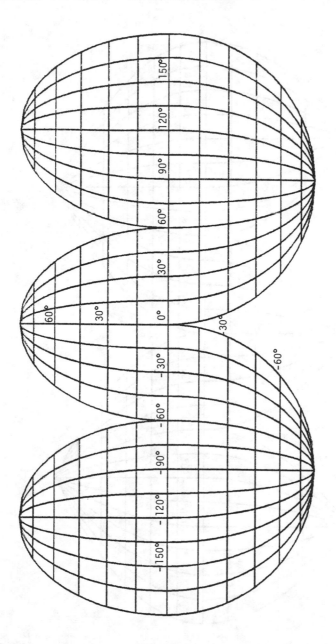

Figure 4.13.2   Interrupted Mollweide Projection

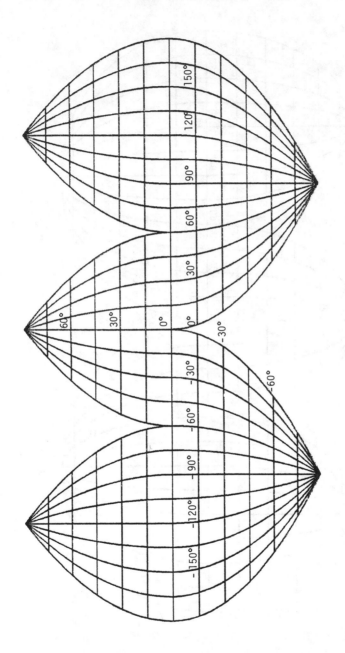

Figure 4.13.3 Interrupted Parabolic Projection

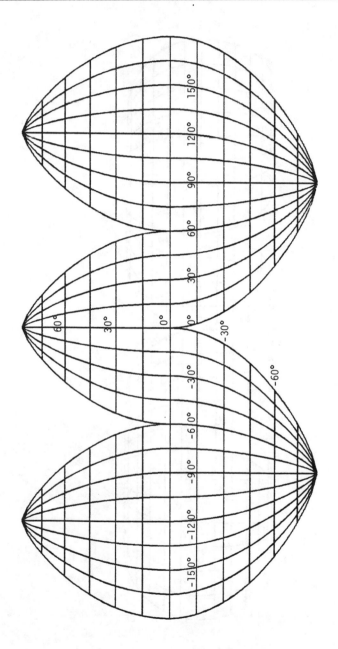

Figure 4.13.4 Interrupted Eumorphic Projection

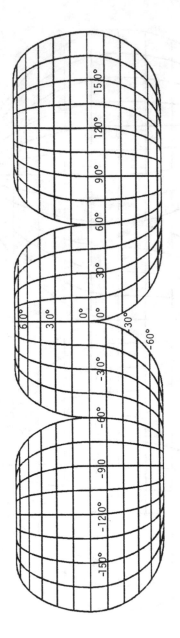

Figure 4.13.5 Interrupted Eckert Projection

undue hardships.   The breaks are chosen to appear in regions of little interest in order to better represent regions of greater interest.   For example, if a continuous map of the world's oceans is desired, the breaks would appear on the land masses.   If, however, land masses were of primary interest, the breaks would be in the oceans.

PROBLEMS

Note: Unless otherwise indicated, take the radius of the authalic sphere to be 6,371,000 meters.

4.1   Find the authalic radius for the IUGG spheroid.

4.2   Find the authalic latitudes for geodetic latitudes of 30°, 45°, and 60° for the IUGG spheroid.

4.3   For a conical equal area projection with one standard parallel, $\lambda_0 = 45°$, $\phi_0 = 30°$, S:1" $\equiv$ 10,000 meters.   For $\phi = 35°$, $\lambda = 50°$, find the cartesian coordinates.

4.4   For conical equal area projection with two standard parallels, $\lambda_0 = 60°$, $\phi_1 = 30°$, $\phi_2 = 40°$.   S:1" $\equiv$ 10,000 meters.   Find the cartesian coordinates when $\phi = 35°$ and $\lambda = 55°$.

4.5   Derive the inverse transformation for the conical equal area projection with two standard parallels.

4.6   Given X = 3", Y = 2" on the conical equal area projection with one standard parallel, $\lambda_0 = 45°$, $\phi_0 = 30°$.   S:1" $\equiv$ 10,000 meters.   Find the inverse point on the authalic sphere.

4.7   Given an azimuthal equal area polar projection with $\lambda = 0°$. S:1" $\equiv$ 20,000 meters $\lambda = 15°$, $\phi = 80°$.   Find the cartesian plotting coordinates.

4.8   On an azimuthal equal area projection with $\lambda_0 = 0°$, X = 5", Y = 5"; S:1" $\equiv$ 20,000 meters.   Find the geographic coordinates on the authalic sphere.

4.9   Derive the inverse transformation equations for the azimuthal equal area oblique projection.

4.10 Derive the inverse transformation equations for the azimuthal equal area equatorial projection.

4.11 Given an azimuthal equal area oblique projection with $\lambda_0 = 0°$, $\phi_0 = 45°$, S:1" $\equiv$ 50,000 meters. $\lambda = 10°$ and $\phi = 50°$.   Find X and Y.

4.12 $\lambda = 0°$ and $\phi = 30°$ on an azimuthal equal area equatorial projection. $\lambda_0 = 10°$ and S:1" $\equiv$ 50,000 meters.   Find the cartesian plotting coordinates.

4.13 On the Bonne projection, $\lambda_0 = 0°$ and $\phi_0 = 30°$. S:1" $\equiv$ 20,000 meters, $\lambda = 10°$ and $\phi = 35°$. Find the cartesian plotting coordinates.

4.14 On the Bonne projection, $\lambda_0 = 0°$ and $\phi_0 = 30°$. X = 2" and Y = 1". S:1" $\equiv$ 20,000 meters. Find the authalic coordinates of the inverse transformation.

4.15 On the cylindrical equal area projection $\lambda_0 = 10°$ and S:1" $\equiv$ 100,000 meters. $\lambda = 35°$ and $\phi = 20°$. Find the cartesian plotting coordinates.

4.16 Given a cylindrical equal area projection with $\lambda_0 = 10°$ and S:1" $\equiv$ 100,000 meters. X = -5", and Y = 2". Find the authalic geographical coordinates.

4.17 On a sinusoidal projection, $\lambda_0 = 10°$ and S:1" $\equiv$ 100,000 meters. $\lambda = 35°$ and $\phi = 20°$. Find the cartesian plotting coodinates.

4.18 On a sinusoidal projection, $\lambda_0 = 10°$ and S:1" $\equiv$ 100,000 meters. X = -5", and Y = 2". Find the authalic geographical coordinates.

4.19 Given a Mollweide projection where $\lambda_0 = 10°$ and S:1" 100,000 meters. $\lambda = 35°$ and $\phi = 20°$. Find the cartesian plotting coordinates. Approximate by means of Figure 4.7.2.

4.20 On a Mollweide projection, $\lambda_0 = 10°$ and S:1" $\equiv$ 100,000 meters. X = -5", and Y = 2". Find the authalic geographical coordinates.

4.21 Given a parabolic projection with $\lambda_0 = 10°$ and S:1" $\equiv$ 100,000 meters. $\lambda = 35°$ and $\phi = 20°$. Find the cartesian plotting coordinates.

4.22 Given a parabolic projection with $\lambda_0 = 10°$ and S:1" $\equiv$ 100,000 meters. X = -5" and Y = 2". Find the authalic geographical coordinates.

4.23 On a Hammer-Aitoff projection with $\lambda_0 = 10°$ and S:1" $\equiv$ 100,000 meters, $\lambda = 35°$ and $\phi = 20°$. What are the cartesian plotting coordinates?

4.24 Given a Hammer-Aitoff projection with S:1" $\equiv$ 100,000 meters, and $\lambda_0 = 10°$. X = -5" and Y = 2". What are the authalic geographical coordinates?

4.25 Derive the inverse Hammer-Aitoff transformation.

4.26 Given the Werner projection with S:1" $\equiv$ 50,000 meters, and $\phi = 30°$ and $\lambda = 30°$, find the cartesian plotting coordinates when $\lambda_0 = 20°$.

4.27 Given the Eckart 4 projection with S:1" $\equiv$ 50,000 meters, and $\phi = 30°$ and $\lambda = 30°$, find the cartesian plotting coordinates when $\lambda_0 = 20°$.

5

Conformal Projections

Conformal projections are those projections which locally maintain the shape of an area on the earth during the transformation to the mapping surface. One important aspect of this is that the orthogonal system of parallels and meridians on the spheroid or sphere appear as an orthogonal system on the map.

The process of conformal transformation may be accomplished in two ways. One of the ways is to transform from the spheroid onto a fictitious conformal sphere, and then apply the simpler spherical formulas to transform from the conformal sphere to the plane, cone, or cylinder. The second way is to accomplish a brute force transformation from the spheroid directly to the mapping surface. Both of these approaches are considered. The results from either approach will be similar.

The conformal projections to be considered are the Mercator, the Lambert conformal, and the stereographic. Three variations of the Mercator are discussed: the regular, the oblique, and the transverse. Lambert conformal projections are represented by one, and two standard parallel cases. The polar, equatorial, and oblique versions of the stereographic are derived. These three projections: the Mercator, the Lambert conformal, and the stereographic permit a global scheme of mapping the earth in sections. The stereographic is best suited for polar regions. The Lambert conformal is best at mid-latitudes and the Mercator is best in the equatorial regions.

All of the conformal projections are characterized by the conformal relation introduced in Section 2.6. The fundamental quantities of the figure of the earth and the mapping surface will be related by

$$\frac{E'}{e} = \frac{G'}{g}.$$

In this equation, the capital letters refer to the mapping surface, and the small letters, to the chosen figure of the earth.

The last section of this chapter presents an important application of conformal projection. This involves their use in state plane coordinate systems. A set of approximate equations are given for the direct and inverse transformations using the Lambert conformal and transverse Mercator projections.

5.1 The Conformal Sphere [22], [25], [26]

The process of producing a conformal mapping of the earth onto a developable surface is aided by the fact that a succession of conformal transformations yields a conformal image of the original area on the final surface. It is shown that it is possible to project the spheroid conformally onto a sphere of radius $R_c = \sqrt{R_p R_m}$, and maintain the quality of conformality for the subsequent transformation to the developable surface.

If one goes through the expansions of this double transformation, and then goes through the expansions for the direct transformation, he will observe that the results are similar, but not exactly the same. The two approaches differ in higher order terms. Since these higher order terms contain powers of the eccentricity, e, the numerical difference is negligible. The process of expansion will not be attempted in this volume. The equations in this section will be derived in a form convenient for evaluation on a computer.

Once the transformation from the spheroidal earth to the conformal sphere is complete, the formulas of trigonometry can be applied to transform from the conformal sphere to the mapping surface.

From Eq. (2.3.15), the first fundamental form of the spheroid is

$$(ds)^2 = R_m^2 (d\phi)^2 + R_p^2 \cos^2 \phi (d\lambda)^2$$

with fundamental quantities

$$e = R_m^2$$

$$g = R_p^2 \cos^2 \phi. \tag{5.1.1}$$

The first fundamental form of the sphere is, from Eq. (2.3.14)

$$(ds)^2 = R_c^2 (d\Phi)^2 + R_c^2 \cos^2 \Phi (d\Lambda)^2$$

with fundamental quantities

$$\left. \begin{array}{l} E' = R_c^2 \\[2mm] G' = R_c^2 \cos^2 \Phi \end{array} \right\} \qquad . \tag{5.1.2}$$

For this conformal sphere, the conformal latitude and longitude are defined to be Φ, and Λ, respectively. The radius of the conformal sphere is R_c.

The conditions are applied that the conformal spherical latitude is a function of spheroidal geodetic latitude only, and conformal longitude is a linear function of spheroidal longitude. Mathematically, this is stated as

$$\left. \begin{array}{l} \Phi = \Phi(\phi) \\[3mm] \Lambda = c\lambda + c_1 \end{array} \right\} \qquad . \tag{5.1.3}$$

Applying the fundamental transformation matrix Eq. (2.7.11) to Eqs. (5.1.2) and (5.1.3)

$$\left. \begin{array}{l} E = \left(\dfrac{\partial \Phi}{\partial \phi}\right)^2 R_c^2 \\[4mm] G = c^2 R_c^2 \cos^2 \phi \end{array} \right\} \qquad . \tag{5.1.4}$$

The condition for conformality, as given by (2.8.2), and applied to the two corresponding orthogonal parametric systems is

$$\frac{E}{e} = \frac{G}{g} = m^2 \tag{5.1.5}$$

where m^2 is a constant.

The fundamental transformation matrix Eq. (2.7.11) gives, with the aid of Eq. (5.1.5)

$$
\left.
\begin{aligned}
E &= \left(\frac{\partial \Phi}{\partial \phi}\right)^2 E' + \left(\frac{\partial \Lambda}{\partial \phi}\right)^2 G' = em^2 \\[2ex]
0 &= \left(\frac{\partial \Phi}{\partial \phi}\right)\left(\frac{\partial \Phi}{\partial \lambda}\right) E' + \left(\frac{\partial \Lambda}{\partial \phi}\right)\left(\frac{\partial \Lambda}{\partial \lambda}\right) G' \\[2ex]
G &= \left(\frac{\partial \Phi}{\partial \lambda}\right)^2 E' + \left(\frac{\partial \Lambda}{\partial \lambda}\right)^2 G' = gm^2
\end{aligned}
\right\} \quad . \tag{5.1.6}
$$

From Eq. (5.1.3) we obtain

$$\frac{\partial \Phi}{\partial \lambda} = \frac{\partial \Lambda}{\partial \phi} = 0 \ . \tag{5.1.7}$$

Substitute Eq. (5.1.7) into Eq. (5.1.6).

$$
\left.
\begin{aligned}
E &= \left(\frac{\partial \Phi}{\partial \phi}\right)^2 E' = em^2 \\[4ex]
G &= \left(\frac{\partial \Lambda}{\partial \lambda}\right)^2 G' = gm^2
\end{aligned}
\right\} \quad . \tag{5.1.8}
$$

Write Eq. (5.1.8) as a proportion to eliminate m^2. This is a form of the condition of conformality.

$$\frac{\left(\frac{\partial \Phi}{\partial \phi}\right)^2}{e} E' = \frac{\left(\frac{\partial \Lambda}{\partial \lambda}\right)^2}{g} G' \ . \tag{5.1.9}$$

Substitute Eqs. (5.1.1), (5.1.2), and the partial derivative of Eq. (5.1.3) into Eq. (5.1.9).

$$\frac{1}{R_m^2} \left(\frac{\partial \Phi}{\partial \phi}\right)^2 R_c^2 = \frac{c^2 R_c^2 \cos^2 \Phi}{R_p^2 \cos^2 \phi} = m^2 \qquad (5.1.10)$$

Convert Eq. (5.1.10) into an ordinary differential equation, and take the square root. Then, by separating the variables.

$$\frac{d\Phi}{\cos \Phi} = \frac{c R_m}{R_p \cos \phi} \, d\phi. \qquad (5.1.11)$$

Substitute Eqs. (3.2.9) and (3.2.16) into Eq. (5.1.11) to obtain

$$\frac{d\Phi}{\cos \Phi} = \frac{\dfrac{ca(1 - e^2)}{(1 - e^2 \sin^2 \phi)^{3/2}}}{\dfrac{a \cos \phi}{(1 - e^2 \sin \phi)^{1/2}}} \, d\phi$$

$$= \frac{c(1 - e^2)}{(1 - e^2 \sin^2 \phi)\cos \phi} \cdot d\phi$$

The solution of this differential equation is

$$\ln \tan \left(\frac{\pi}{4} + \frac{\Phi}{2}\right) = c \ln \tan \left(\frac{\pi}{4} + \frac{\phi}{2}\right)\left(\frac{1 - e \sin \phi}{1 + e \sin \phi}\right)^{e/2} + K. \qquad (5.1.12)$$

The constant K is removed by requiring that Φ and ϕ are coincidently equal to zero. Thus, from (5.1.12)

$$\tan \left(\frac{\pi}{4} + \frac{\Phi}{2}\right) = \left\{\tan \left(\frac{\pi}{4} + \frac{\phi}{2}\right)\left(\frac{1 - e \sin \phi}{1 + e \sin \phi}\right)^{e/2}\right\}^c. \qquad (5.1.13)$$

Note that this integral was encountered before, in Section 3.3, for the loxodromic curve on the spheroid.

In order that Λ and λ coincide at zero, from the second of Eqs. (5.1.3), $c_1 = 0$, and

$$\Lambda = c\lambda. \qquad (5.1.14)$$

It remains to find the value of the constant c for the particular transformation from the spheroid to the conformal sphere.

Consider a Taylor's series expansion of the constant m^2 about the origin. By origin we mean in this development, the latitude selected as the origin of the map. Recall that the partial derivatives of m^2 with respect to λ are zero. Then,

$$m^2 = m_0^2 + \left(\frac{\partial m^2}{\partial \phi}\right)_0 \Delta\phi + \frac{1}{2}\left(\frac{\partial^2 m^2}{\partial \phi^2}\right)_0 (\Delta\phi)^2 + \dots \qquad (5.1.15)$$

Also from eq. (5.1.10)

$$m = \frac{cR_c \cos \Phi}{R_p \cos \phi}. \qquad (5.1.16)$$

At the origin of the map, $m_0 = 1$, by definition of the conformal projection. This aspect of map projections will be explored in Chapter 7 on the theory of distortion. Considering Eq. (5.1.6) at the origin.

$$m_0 = 1 = \frac{cR_c \cos \Phi_0}{R_p \cos \phi_0}. \qquad (5.1.17)$$

Let

$$\frac{\partial m}{\partial \phi} = cR_c\left\{\frac{\partial}{\partial \phi}\left[\frac{\cos \Phi}{R_p \cos \phi}\right]\right\} = 0. \qquad (5.1.18)$$

Taking the derivative of the portion of Eq. (5.1.6) in brackets, and substituting Eq. (3.2.9) in this gives the following.

$$-\frac{\sin \phi}{R_p \cos \phi}\frac{\partial \Phi}{\partial \phi} = \frac{\cos \Phi}{(R_p \cos \phi)^2}\left(-R_p \sin \phi + \cos \phi \frac{\partial R_p}{\partial \phi}\right) = 0$$

$$\sin \Phi \frac{\partial \Phi}{\partial \phi} = - \frac{\cos \Phi}{R_p \cos \phi} \left[-R_p \sin \phi + \frac{\cos \phi \, ae^2 \sin \phi \cos \phi}{(1 - e^2 \sin^2 \phi)^{3/2}}\right]$$

$$= - \frac{\cos \Phi \sin \phi}{R_p \cos \phi} \left[-R_p - \frac{ae^2 \cos^2 \phi}{(1 - e^2 \sin^2 \phi)^{3/2}}\right]$$

$$= \frac{\cos \Phi \sin \phi}{R_p \cos \phi} \left[\frac{a(1 - e^2 \sin^2 \phi) - ae^2 \cos^2 \phi}{(1 - e^2 \sin^2 \phi)^{3/2}}\right]$$

$$= \frac{\cos \Phi \sin \phi}{R_p \cos \phi} \left[\frac{a(1 - e^2)}{(1 - e^2 \sin^2 \phi)^{3/2}}\right]. \tag{5.1.19}$$

Substitute Eq. (3.2.16) into Eq. (5.1.19).

$$\sin \Phi \frac{\partial \Phi}{\partial \phi} = \cos \Phi \frac{R_m \sin \phi}{R_p \cos \phi}. \tag{5.1.20}$$

Evaluate Eq. (5.1.20) at the origin of the map.

$$\sin \Phi_0 \left(\frac{\partial \Phi}{\partial \phi}\right)_0 = \cos \Phi_0 \frac{R_{mo} \sin \phi_0}{R_{po} \cos \phi_0}. \tag{5.1.21}$$

Substitute Eq. (5.1.10) into Eq. (5.1.21).

$$\frac{\sin \Phi_0 \cos \Phi_0 \, cR_{mo}}{R_{po} \cos \phi_0} = \frac{\cos \Phi_0 \, R_{mo} \sin \phi_0}{R_{po} \cos \phi_0}$$

$$\sin \phi_0 = c \, \sin \Phi_0. \tag{5.1.22}$$

The next step is to obtain the second partial derivative of m, and equate this to zero. This is accomplished by using Eq. (5.1.23) to obtain

$$\tan \phi_0 = \frac{R_{po}}{R_{mo}} \tan \phi_0. \tag{5.1.23}$$

From Eq. (5.1.22)

$$\sin \Phi_0 = \frac{\sin \phi_0}{c} \qquad\qquad\qquad (5.1.24)$$

$$\cos \Phi_0 = \sqrt{1 - \frac{\sin^2 \phi_0}{c^2}}. \qquad\qquad\qquad (5.1.25)$$

Substitute Eqs. (5.1.24) and (5.1.25) into Eq. (5.1.23). Also, substitute Eq. (3.2.9) and Eq. (3.2.16).

$$\frac{\sin \phi_0}{\cos \phi_0} = \sqrt{\frac{R_{po}}{R_{mo}}} \frac{\sin \Phi_0}{\cos \Phi_0} = \sqrt{\frac{\dfrac{a}{(1 - e^2 \sin^2 \phi_0)^{1/2}}}{\dfrac{a(1 - e^2)}{(1 - e^2 \sin^2 \phi_0)^{3/2}}}} \cdot \frac{\dfrac{\sin \phi_0}{c}}{\sqrt{1 - \dfrac{\sin^2 \phi_0}{c}}}$$

$$\frac{1}{\cos \phi_0} = \sqrt{\frac{1 - e^2 \sin^2 \phi_0}{1 - e^2}} \frac{1}{c^2 - \sin^2 \phi_0}$$

$$\frac{1}{\cos^2 \phi_0} = \frac{1 - e^2 \sin^2 \phi_0}{1 - e^2} \cdot \frac{1}{c^2 - \sin^2 \phi_0}$$

$$\cos^2 \phi_0 = \frac{(1 - e^2)(c^2 - \sin^2 \phi_0)}{1 - e^2 \sin^2 \phi_0}$$

$$\cos^2 \phi_0 (1 - e^2 \sin^2 \phi_0) = (1 - e^2)(c^2 - \sin^2 \phi_0)$$

$$\cos^2 \phi_0 - e^2 \cos^2 \phi_0 \sin^2 \phi_0 = (1 - e^2)c^2 - \sin^2 \phi_0 + e^2 \sin^2 \phi_0$$

$$\cos^2 \phi_0 + \sin^2 \phi_0 - e^2 \sin^2 \phi_0 (1 + \cos^2 \phi_0) = c^2 (1 - e^2)$$

$$1 - e^2 (1 - \cos^2 \phi_0)(1 + \cos^2 \phi_0) = c^2 (1 - e^2)$$

$$1 + e^2 (\cos^4 \phi_0 - 1) = c^2 (1 - e^2)$$

$$c^2 = \frac{1 - e^2 + e^2 \cos^4 \phi_0}{1 - e^2}$$

$$c = (1 + \frac{e^2 \cos^4 \phi_0}{1 - e^2})^{1/2}. \tag{5.1.26}$$

The radius of the conformal sphere can be found from Eq. (5.1.23)

$$\frac{\sin \phi_0}{\cos \phi_0} = \sqrt{\frac{R_{po}}{R_{mo}}} \cdot \frac{\sin \Phi_0}{\cos \Phi_0}. \tag{5.1.27}$$

From Eq. (5.1.24)

$$\frac{c \sin \Phi_0}{\cos \phi_0} = \sqrt{\frac{R_{po}}{R_{mo}}} \frac{\sin \Phi_0}{\cos \Phi_0} \quad , \quad c = \frac{\cos \phi_0}{\cos \Phi_0} \left(\frac{R_{po}}{R_{mo}}\right)^{1/2}. \tag{5.1.28}$$

Eliminate c between Eq. (5.1.17) and Eq. (5.1.28)

$$1 = \frac{R_c}{\sqrt{R_{po}R_{mo}}}$$

$$R_c = \sqrt{R_{po}R_{mo}}. \tag{5.1.29}$$

As an example, let ϕ_0 = 45°. From Table 3.2.1 for the WGS-72 spheroid, R_{po} = 6,388,836 meters and R_{mo} = 6,367,380 meters. Then,

$$R_c = \sqrt{R_{po}R_{mo}} = \sqrt{(6,388,836)(6,367,380)}$$
$$= 6,478,099 \text{ meters}.$$

Equations (5.1.13), (5.1.14), and (5.1.26) can be used to convert from the spheroidal earth to a conformally equivalent sphere, with a radius given by Eq. (5.1.29). Once this is done, the conformal projection from the conformal sphere is relatively easy. Table 5.1.1 gives the conformal latitude in terms of geodetic latitude for the WGS-72 ellipsoid when ϕ_0 is arbitrarily chosen as 0°. Note that, unlike the development for authalic latitude the conformal sphere depends on a particular choice of origin, ϕ_0. Note also that Eq. (5.1.29), or the radius of the conformal sphere is also dependent on this origin. Thus, the radius of the conformal sphere contracts or expands as the choice of the origin dictates.

Table 5.1.1
Conformal Latitude as a Function of Geodetic
Latitude for the WGS-72 Spheroid

Geodetic Latitude	Conformal Latitude
0.	.0000
5.	4.9836
10.	9.9680
15.	14.9538
20.	19.9418
25.	24.9325
30.	29.9264
35.	34.9238
40.	39.9248
45.	44.9296
50.	49.9378
55.	54.9491
60.	59.9628
65.	64.9781
70.	69.9938

Origin at $\phi_0 = 0°$

5.2 Mercator Projection [8], [22], [24], [25], [26]

The Mercator projection, devised in 1569 by Gerhard Kramer, whose Latin name was Mercator, is the classic of modern map projections. It was derived as an aid to navigation in the initial days of the age of ocean exploration, and has continued its utility through the age of space exploration. The regular, or equatorial, Mercator projection, with its areas of lesser distortion north and south of the equator, and including the major maritime trade routes was and is a natural vehicle for ocean navigation. Transverse Mercator projections, with the Lambert conformal, are the backbone of the quadrangle system for topographic surveying. Oblique Mercator projections, with the line of zero distortion along the nominal satellite re-entry footprint have been used in the recovery charts for the Mercury, Gemini, and Apollo missions.

All three of the Mercator variations, the regular, the oblique, and the transverse are here considered in terms of a double transformation, that is from the spheroid to the conformal sphere, and then to the mapping surface, and the equatorial variation in terms of a direct transformation of the spheroid to the map.

The Mercator projection entails, in both approaches, a transformation from the spheroid to a cylinder. The Mercator projection can be considered in terms of a semi-graphical technique. One can, with extreme patience, construct a Mercator projection by a graphical means. In fact, before the development of calculus, Mercator did just that. The objection to this is that there is a varying projection point. Calculations are needed to locate the point of emanation of the projection ray. This is shown in Figure 5.2.1. For each and every latitude, a different point is needed as the origin of an interior ray which intersect both the surface of the spheroid (or conformal sphere) and the developable surface, the cylinder. Thus, the reasonable approach is to use a mathematical method.

The regular Mercator projection will be developed first for the conformal sphere. While the rotation formulas of Section 2.10 can be applied to produce the oblique and transverse cases, a spherical trigonometric approach is used in this text.

For the regular Mercator projection, let the cartesian mapping coordinates be given by the functional relations.

$$\left. \begin{array}{l} x = x(\lambda) \\ y = y(\phi) \end{array} \right\} \quad . \tag{5.2.1}$$

In particular, the first function is taken as a linear combination

$$x = aS(\lambda - \lambda_0) \tag{5.2.2}$$

where a is the radius of the conformal sphere, and S is the scale factor.

The differential forms of the second of Eq. (5.2.1), and (5.2.2) are

$$\left. \begin{array}{l} dx = aS \, d\lambda \\ dy = \dfrac{dy}{d\phi} \, d\phi \end{array} \right\} \quad . \tag{5.2.3}$$

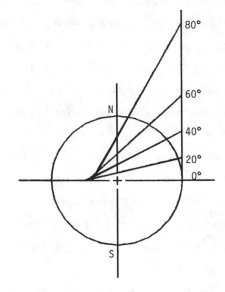

Figure 5.2.1   The Varying Projection Point of the Equatorial Mercator Projection

The first fundamental form of the plane is

$$(ds)^2 = (dy^2) + (dx)^2.$$ (5.2.4)

Substitute Eqs. (5.2.3) into (5.2.4).

$$(ds)^2 = (\frac{dy}{d\phi})^2(d\phi)^2 + a^2 S^2(d\lambda)^2.$$ (5.2.5)

The first fundamental quantities of Eq. (5.2.5) are

$$E = (\frac{dy}{d\phi})^2$$

$$\left.\begin{array}{c} \\ \\ \\ \end{array}\right\} \quad . \qquad\qquad (5.2.6)$$

$$G = a^2 S^2$$

The first fundamental form of the conformal sphere is

$$(ds)^2 = a^2 (d\phi)^2 + a^2 \cos^2 \phi (d\lambda)^2 . \qquad\qquad (5.2.7)$$

The first fundamental quantities, from Eq. (5.2.7), are

$$e = a^2$$

$$\left.\begin{array}{c} \\ \\ \\ \end{array}\right\} \quad . \qquad\qquad (5.2.8)$$

$$g = a^2 \cos^2 \phi$$

For the orthogonal systems of the plane and the conformal sphere, the relation of conformality, from Eq. (2.8.2) is

$$\frac{E}{e} = \frac{G}{g} . \qquad\qquad (5.2.9)$$

Substituting Eqs. (5.2.6) and (5.2.8) into Eq. (5.2.9), simplifying, and integrating

$$\frac{a^2 S^2}{a^2 \cos^2 \phi} = \frac{(\frac{dy}{d\phi})^2}{a^2}$$

$$\frac{dy}{d\phi} = \frac{aS}{\cos \phi}$$

$$y = a \int \frac{S \, d\phi}{\cos \phi} = aS \ln \tan (\frac{\pi}{4} + \frac{\phi}{2}) + c . \qquad\qquad (5.2.10)$$

In Eq. (5.2.10), choose c such that $y = 0$ when $\phi = 0$. Then, $c = 0$.

$$y = aS \ln \tan (\frac{\pi}{4} + \frac{\phi}{2}) . \qquad\qquad (5.2.11)$$

Equations (5.2.2) and (5.2.11) provide the basis for the transformation from the conformal sphere to the cylindrical plotting surface. They are now used with a specific spherical trigonometric approach to obtain the oblique

and transverse Mercator variations.   Figure 5.2.2 is used to define the
necessary spherical trigonometry.   In this figure point $P(\phi_p, \lambda_p)$ is the pole
of the desired projeciton, and $Q(\phi, \lambda)$ is an arbitrary point.   The procedure is
to set up relations for an arbitrary oblique transformation, and then
particularize for the transverse case.

The first step is to set up the relations between ϕ and λ in the
equatorial system, and auxiliary variables h' and α in the system with the
pole at O.   The variables involved in the transformation are indicated on the
figure.   From the spherical triangle NQP, we can obtain the following
equations, with $\Delta\lambda_p = \lambda - \lambda_p$

$$\cos h' = \sin \phi_p \sin \phi + \cos \phi_p \cos \phi \cos \Delta\lambda_p \qquad (5.2.12)$$

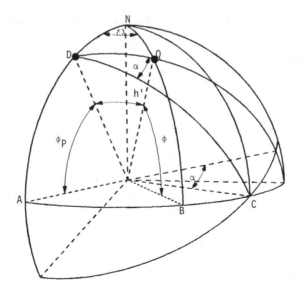

Figure 5.2.2 Geometry for the Development of the Oblique and Transverse
          Mercator Projections

$$\sin h' \cos \alpha = \cos \phi \sin \Delta\lambda_p \qquad (5.2.13)$$

$$\sin h' \sin \alpha = \cos \phi_p \sin \phi - \sin \phi_p \cos \phi \cos \Delta\lambda_p \qquad (5.2.14)$$

Writing Eq. (5.2.11) in terms of the colatitude $z = \frac{\pi}{2} - \phi$

$$y = a \ln \cot z/2 \qquad (5.2.15)$$

The corresponding Mercator forms for Eqs. (5.2.2) and (5.2.15) in the system with the pole at P are

$$x = a\alpha$$

$$y = a \ln \cot \frac{h'}{2} \qquad (5.2.16)$$

However, since it is desired that the great circle PC' be true length, it is necessary to interchange the values of x and y in Eqs. (5.2.16) to obtain

$$x = a \ln \cot \frac{h'}{2} \qquad (5.2.17)$$

$$y = a\alpha \qquad (5.2.18)$$

Applying a trigonometric identity in Eq. (5.2.17), we obtain

$$x = \frac{a}{2} \ln \left(\frac{1 + \cos h'}{1 - \cos h'}\right) \qquad (5.2.19)$$

Substitute Eq. (5.2.12) into Eq. (5.2.19)

$$x = \frac{a}{2} \ln \left(\frac{1 + \sin \phi_p \sin \phi + \cos \phi_p \cos \phi \cos \Delta\lambda_p}{1 - \sin \phi_p \sin \phi - \cos \phi_p \cos \phi \cos \Delta\lambda_p}\right) \qquad (5.2.20)$$

Dividing Eq. (5.2.14) by Eq. (5.2.13), we obtain the following.

$$\tan \alpha = \frac{\cos \phi_p \sin \phi - \sin \phi_p \cos \phi \cos \Delta\lambda_p}{\cos \phi \sin \Delta\lambda_p} \qquad (5.2.21)$$

Substituting the arctangent of (5.2.21) into Eq. (5.2.17)

$$y = a \tan^{-1} \left(\frac{\cos \phi_p \sin \phi - \sin \phi_p \cos \phi \cos \Delta\lambda_p}{\cos \phi \sin \Delta\lambda_p}\right) \qquad (5.2.22)$$

With the inclusion of the scale factor, S, Eqs. (5.2.20) and (5.2.22) give the plotting equations for the oblique Mercator projection. To obtain the transverse Mercator projection, it is necessary to substitute $\phi_p = 0°$ into

Eqs. (5.2.20) and (5.2.22). It is also necessary to transform the center of coordinates by replacing $\Delta\lambda$ by $\Delta\lambda - \frac{\pi}{2}$, where $\Delta\lambda = \lambda - \lambda_0$, and λ_0 is the central meridian. This provides the following equations when the scale factor is included.

$$\left. \begin{aligned} x &= \frac{aS}{2} \ln \left(\frac{1 - \cos\phi \sin \Delta\lambda}{1 + \cos\phi \sin \Delta\lambda}\right) \\ \\ y &= aS \tan^{-1} \left(\frac{\tan\phi}{\cos \Delta\lambda}\right) \end{aligned} \right\} . \qquad (5.2.23)$$

Figures 5.2.3 and 5.2.4 are specimens of the equatorial and transverse Mercator projections, respectively. Plotting tables for these projections are given in Tables 5.2.1 and 5.2.2 respectively. Note that, in the equatorial Mercator, the parallels and meridians are straight lines, intersecting at

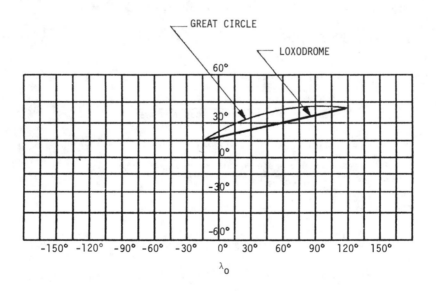

Figure 5.2.3 Regular or Equatorial Mercator Projection

right angles. This means that the convergency of the meridians does not occur, and distortion becomes excessive in a poleward direciton. In fact, the point of the pole is approaching infinity. Thus, the equatorial projection is useless at extremely high and low latitudes. In the transverse Mercator, the central meridian and the equator are the only straight lines. However, curved meridians and parallels intersect orthogonally. For the oblique Mercator, there are no straight meridians or parallels. However, orthogonality is present.

For the regular, or equatorial, Mercator projection, the inverse transformation from cartesian coordinates to conformal geograhic coordinate follow relatively simply from Eqs. (5.2.2) and (5.211)

$$\Delta\lambda = \frac{x}{aS}$$ (5.2.24)

Table 5.2.1
Equatorial Mercator Projection

Latitude Degrees	Longitude Degrees	X Meters	Y Meters
0.	0.	0.000	.000
0.	30.	3.340	.000
0.	60.	6.679	.000
0.	90.	10.019	.000
0.	120.	13.359	.000
0.	150.	16.698	.000
0.	180.	20.038	.000
30.	0.	.000	3.482
30.	30.	3.340	3.482
30.	60.	6.679	3.482
30.	90.	10.019	3.482
30.	120.	13.359	3.482
30.	150.	16.698	3.482
30.	180.	20.038	3.482
60.	0.	.000	8.363
60.	30.	3.340	8.363
60.	60.	6.679	8.363
60.	90.	10.019	8.363
60.	120.	13.359	8.363
60.	150.	16.698	8.363
60.	180.	20.038	8.363

Map Origin $\phi_0 = 90^0$

$\lambda_0 = 0^0$

Table 5.2.2
Transverse Mercator Projection

Latitude Degrees	Longitude Degrees	X Meters	Y Meters
30.	0.	.000	-8.400
30.	15.	2.688	-7.714
30.	30.	4.552	-6.206
30.	45.	5.652	-4.546
30.	60.	6.268	-2.957
30.	75.	6.583	-1.454
30.	90.	6.679	.000
45.	0.	.000	-5.621
45.	15.	1.615	-5.324
45.	30.	2.957	-4.546
45.	45.	3.926	-3.503
45.	60.	4.552	-2.357
45.	75.	4.899	-1.180
45.	90.	5.009	.000
60.	0.	.000	-3.503
60.	15.	.946	-3.360
60.	30.	1.792	-2.957
60.	45.	2.472	-2.357
60.	60.	2.957	-1.629
60.	75.	3.245	- .830
60.	90.	3.339	.000
75.	0.	.000	-1.689
75.	15.	.442	-1.629
75.	30.	.849	-1.454
75.	45.	1.194	-1.180
75.	60.	1.454	- .830
75.	75.	1.615	- .428
75.	90.	1.670	.000
90.	0.	0.038	.000
90.	15.	.037	.000
90.	30.	.037	.000
90.	45.	.037	.000
90.	60.	.037	.000
90.	75.	.037	.000
90.	90.	.037	.000

Map Origin: $\phi_0 = 0^0$

Central meridian: $\lambda_0 = 0^0$

$$\phi = 2[\tan^{-1}(\varepsilon^{y/aS}) - \frac{\pi}{4}]$$

where ε is the base of natural logarithms. It is obvious from an inspection of the plotting equations for the oblique and transverse Mercator projections that their inverses are not so simple. To partially overcome this difficulty,

Section 5.4 gives an approximate solution in the case of the inverse transformation for the transverse Mercator. This approximation has an accuracy compatible with practical computing requirements.

In Figure 5.2.3, the loxodrome (or rhumbline) and the great circle are portrayed on a equatorial Mercator projection. The loxodrome is a line which intersects successive meridians at the same azimuth, or bearing angle. On the Mercator projection the loxodrome is a straight line, and the great circle (or geodesic) is a curved line. The gnomonic projection of Section 6.1 has the reverse of this situation. As is seen, the gnomonic projection has great circles as straight lines, and the loxodromes are curved lines. Thus, by using the Mercator and the gnomonic projections together, one can build a series of bearings which approximates, piecewise, a great circle route. This combines ease of navigation with an approximation to the shortest distance

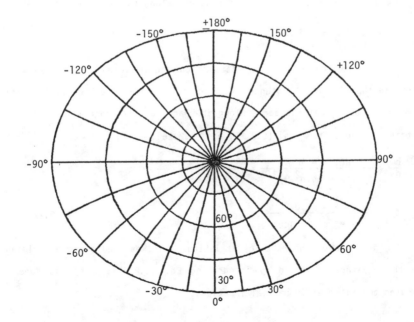

Figure 5.2.4 Transverse Mercator Projection

between two arbitrary points. This method is used for both maritime and aerial navigation.

The Mercator projection, as mentioned above, can also be derived by a direct transformation from the spheroid to the plotting surface. This is now done to compare the labor in these two approaches.

From Eq. (3.3.1), an element of distance along a parallel of the spheroid is

$$dp = \frac{a \cos \phi \, d\lambda}{(1 - e^2 \sin^2 \phi)^{1/2}}.$$

The infinitesimal distance along the parallel of the map is $ad\lambda$. Thus, the scale along the parallel is

$$\frac{dp}{ad\lambda} = \frac{\cos \phi}{(1 - e^2 \sin^2 \phi)^{1/2}}. \qquad (5.2.25)$$

From Eq. (3.2.16), an element of distance along the meridian is given by

$$dm = \frac{a(1 - e^2)d\phi}{(1 - e^2 \sin^2 \phi)^{3/2}}. \qquad (5.2.26)$$

Let dy be the element of distance on the meridian of the map which represents the elemental distance dm along the meridianal ellipse. The ratio of dm to dy must equal the scale along the parallel, if conformality is to be maintained. Thus, from Eqs. (5.2.25) and (5.2.26).

$$\frac{dm}{dy} = \frac{a(1 - e^2)d\phi}{dy(1 - e^2 \sin^2 \phi)^{3/2}} = \frac{\cos \phi}{(1 - e^2 \sin^2 \phi)^{1/2}}$$

$$dy = \frac{a(1 - e^2)d\phi}{(1 - e^2 \sin^2 \phi)\cos \phi}. \qquad (5.2.27)$$

The distance of the parallel of latitude, as measured along a meridian, from the equator, is found by integrating Eq. (5.2.27). This is done by expanding the integral in partial fractions.

$$y = \int_0^\phi \frac{a(1 - e^2)d\phi}{(1 - e^2 \sin^2 \phi)\cos \phi}$$

$$= a\{\int_0^\phi \frac{d\phi}{\cos \phi} + \frac{e}{2} \int_0^\phi \frac{-e \cos \phi \, d\phi}{1 - e \sin \phi} - \frac{e}{2} \int_0^\phi \frac{e \cos \phi \, d\phi}{1 + e \sin \phi}\}$$

$$= a\{\int_0^\phi \frac{d\phi}{\sin(\frac{\pi}{2} + \phi)} + \frac{e}{2} \int_0^\phi \frac{-e \cos \phi \, d\phi}{1 - e \sin \phi} - \frac{e}{2} \int_0^\phi \frac{e \cos \phi \, d\phi}{1 - e \sin \phi}\}$$

$$= a\{\int_0^\phi \frac{\cos(\frac{\pi}{4} + \frac{\phi}{2})}{\sin(\frac{\pi}{4} + \frac{\phi}{2})} \frac{d\phi}{2} - \int_0^\phi \frac{\sin(\frac{\pi}{4} + \frac{\phi}{2})}{\cos(\frac{\pi}{4} + \frac{\phi}{2})} \frac{d\phi}{2}$$

$$+ \frac{e}{2} \int_0^\phi \frac{-e \cos \phi \, d\phi}{1 - e \sin \phi} - \frac{e}{2} \int_0^\phi \frac{e \cos \phi \, d\phi}{1 + e \sin \phi}\}$$

$$y = a\{\ln[\sin(\frac{\pi}{4} + \frac{\phi}{2})] - \ln[\cos(\frac{\pi}{4} + \frac{\phi}{2})]$$

$$+ \frac{e}{2} \ln(1 - e \sin \phi) - \frac{e}{2} \ln(1 + e \sin \phi)]\}$$

$$= a\{\ln[\tan(\frac{\pi}{4} + \frac{\phi}{2})] + \frac{e}{2} \ln[(\frac{1 - e \sin \phi}{1 + e \sin \phi})]$$

$$= a \ln[\tan(\frac{\pi}{4} + \frac{\phi}{2})(\frac{1 - e \sin \phi}{1 + e \sin \phi})^{e/2}\}. \qquad (5.2.28)$$

Thus, we have in Eq. (5.2.28) the same results as in Eq. (5.2.14) if e is set equal to zero.

The distance along the equator can be found from the integral

$$x = a \int_0^\lambda d\lambda$$

$$= a\Delta\lambda. \qquad (5.2.29)$$

The amount of labor in either the direct or the indirect method of transformation is significant. The amount of computer time consumed in evaluating either set of equations is similar. The rotations of Section 2.10

can be applied to Eqs. (5.2.28) and (5.2.29) to obtain approximations to the oblique and transverse Mercator projections.

5.3 Lambert Conformal [22], [25]

The Lambert conformal projection is a projection from the spheroidal earth onto a cone, which serves as the developable surface. This can be done rather simply by transforming from the conformal sphere to the cone. The transformation can also be accomplished directly from the spheroid to the cone. This second approach is followed in this section for Lambert conformal projections with one and two standard parallels. Then, the eccentricity is set equal to zero to accommodate transformations from the conformal sphere.

For the Lambert conformal projection with one standard parallel, the conical mapping surface is tangent to the spheroid at this standard parallel. The axis of the cone coincides with the rotation, or polar axis of the earth. The meridians are straight lines converging at the apex of the cone. One of these meridians is arbitrarily chosen as the central meridian, λ_0. The parallels are a set of concentric circles.

The polar coordinates of a point P are ρ and θ. The cartesian coordinates of this same point are

$$\left. \begin{array}{l} x = \rho \sin \theta \cdot S \\[2mm] y = (\rho_0 - \rho \cos \theta)S \end{array} \right\} \quad . \tag{5.3.1}$$

where ρ_0 is the distance from the apex of the cone to the circle of tangency.

Again, the elemental distance on the spheroid is found from the first fundamental form

$$(ds)^2 = R_m^2 (d\phi)^2 + R_p^2 \cos^2 \phi (d\lambda)^2$$

with fundamental quantities

$$\left. \begin{array}{l} e = R_m^2 \\[3mm] g = R_p^2 \cos^2 \phi \end{array} \right\} \quad . \tag{5.3.2}$$

On the conical surface, the first fundamental form is

$$(ds)^2 = (d\rho)^2 + \rho^2(d\theta)^2$$

with fundamental quantities

$$\left.\begin{array}{l} E' = 1 \\ G' = \rho^2 \end{array}\right\} \cdot \qquad (5.3.3)$$

Two conditions will be imposed. One is that

$$\rho = \rho(\phi) \qquad (5.3.4)$$

and the second is that

$$\theta = c_1\lambda + c_2. \qquad (5.3.5)$$

From Eq. (5.3.5)

$$\frac{\partial\theta}{\partial\lambda} = c_1. \qquad (5.3.6)$$

From the fundamental transformation matrix, Eq. (2.7.11)

$$\left.\begin{array}{l} E = (\frac{\partial\rho}{\partial\phi})^2 E' \\ \\ G = (\frac{\partial\theta}{\partial\lambda})^2 G' \end{array}\right\} \cdot \qquad (5.3.7)$$

Substitute Eq. (5.3.3) into Eq. (5.3.7)

$$\left.\begin{array}{l} E = (\frac{\partial\rho}{\partial\phi})^2 \\ \\ G = (\frac{\partial\theta}{\partial\lambda})^2 \rho^2 \end{array}\right\} \cdot \qquad (5.3.8)$$

Substitute Eq. (5.3.6) into the second Eqs. of (5.3.8).

$$G = c_1^2 \rho^2. \qquad (5.3.9)$$

From the condition of conformality Eq. (5.1.9) for two orthogonal systems.

$$\frac{(\frac{\partial \rho}{\partial \phi})^2}{e} = \frac{(\frac{\partial \theta}{\partial \lambda})^2 \rho^2}{g} = m^2.$$

(5.3.10)

Substitute Eq. (5.3.2) into Eq. (5.3.10).

$$\frac{(\frac{\partial \rho}{\partial \phi})^2}{R_p^2} = \frac{c_1^2 \rho^2}{R_p^2 \cos^2 \phi} = m^2.$$

(5.3.11)

Take the square root of Eq. (5.3.11), and convert the result to an ordinary differential equation

$$\frac{d\rho}{\rho} = - \frac{R_m c_1}{R_p \cos \phi} d\phi.$$

(5.3.12)

The minus sign is chosen since ρ decreases as ϕ increases.

Equation (5.3.12) can be integrated by the method of Section 5.2 for the Mercator projection to obtain

$$\ln \rho = -c_1 \ln\{\tan(\frac{\pi}{4} + \frac{\phi}{2})(\frac{1 - e \sin \phi}{1 + e \sin \phi})^{e/2}\}$$

$$+ \ln c_3$$

$$\rho = c_3 \{\tan(\frac{\pi}{4} - \frac{\phi}{2})(\frac{1 + e \sin \phi}{1 - e \sin \phi})^{e/2}\}^{c_1}.$$

(5.3.13)

The constants c_1, c_2, and c_3 must be evaluated now. First, from Eq. (5.3.5), it is required that $\theta = 0$, when $\lambda = 0$. Thus, $c_2 = 0$.

Next, consider c_3. At the origin of the cartesian coordinate system of the map (ϕ_0, λ_0), the cone is tangent to the spheroid. Thus, similar to the development of Section 1.12

$$\rho_0 = R_{po} \cot \phi_0.$$

(5.3.14)

Evaluate Eq. (5.3.13) at ϕ_0, and equate to eq. (5.3.14).

$$R_{po} \cot \phi_0 = c_3 \{ \tan (\frac{\pi}{4} - \frac{\phi_0}{2})(\frac{1 + e \sin \phi_0}{1 - e \sin \phi_0})^{e/2} \}^{c_1}$$

$$c_3 = \frac{R_{po} \cot \phi_0}{\{ \tan(\frac{\pi}{4} - \frac{\phi_0}{2})(\frac{1 + e \sin \phi_0}{1 - e \sin \phi_0})^{e/2} \}^{c_1}} . \qquad (5.3.15)$$

Finally, from Eq. (5.3.11)

$$m = \frac{c_1 \rho}{R_p \cos \phi} . \qquad (5.3.16)$$

At the origin, as is treated in detail in Chapter 7, $m_0 = 1$. This implies that $(\partial m / \partial \phi)_0 = 0$. Differentiate Eq. (5.3.4), and evaluate this at the origin.

$$\frac{\partial m}{\partial \phi} = c_1 (\frac{\partial \rho}{\partial \phi}) \frac{1}{R_p \cos \phi} + c_1 \rho \frac{R_m \sin \theta}{R_p^2 \cos^2 \phi} = 0$$

$$c_1 (\frac{\partial \rho}{\partial \phi})_0 + \frac{c_1 \rho_0 R_{mo}}{R_{po} \cos \phi_0} = 0 . \qquad (5.3.17)$$

From Eq. (5.3.12) we obtain

$$(\frac{\partial \rho}{\partial \phi_0}) = - \frac{c_1 \rho_0 R_{mo}}{R_{po} \cos \phi_0} . \qquad (5.3.18)$$

Substitute Eq. (5.3.18) into Eq. (5.3.17)

$$- \frac{c_1^2 \rho_0 R_{mo}}{R_{po} \cos \phi_0} + \frac{c_1 \rho_0 R_{mo} \sin \phi_0}{R_{po} \cos \phi_0} = 0$$

$$c_1 = \sin \phi_0 . \qquad (5.3.19)$$

Substitute Eqs. (5.3.19), and (5.3.14), into Eq. (5.3.15)

$$c_3 = \frac{\rho_0}{\{ \tan(\frac{\pi}{4} - \frac{\phi_0}{2})(\frac{1 - e \sin \phi_0}{1 - e \sin \phi_0})^{3/2} \}^{\sin \phi_0}} \qquad (5.3.20)$$

Substitute Eq. (5.3.20) into Eq. (5.3.13).

$$\rho = \rho_0 \left\{ \frac{\tan(\frac{\pi}{4} - \frac{\phi}{2})(\frac{1 + e \sin \phi}{1 - e \sin \phi})^{e/2}}{\tan(\frac{\pi}{4} - \frac{\phi_0}{2})(\frac{1 + e \sin \phi_0}{1 - e \sin \phi_0})^{e/2}} \right\}^{\sin \phi_0} . \qquad (5.3.21)$$

Substitute Eq. (5.3.19) into Eq. (5.3.5), and recall that $c_2 = 0$.

$$\theta = \lambda \sin \phi_0. \qquad (5.3.22)$$

Recall from Section 1.12 that $\sin \phi_0$ is the constant of the cone.

Equations (5.3.21) and (5.3.22), in conjunction with eqs. (5.3.1) and (5.3.14) give the plotting equations in cartesian coordinates. Table 5.3.1 is a plotting table for the Lambert conformal projection with one standard

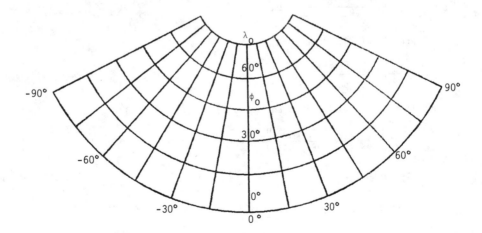

Figure 5.3.1 Lambert Conformal Projection, One Standard Parallel

Table 5.3.1
Lambert Conformal Projection, One Standard Parallel.

Latitude Degrees	Longitude Degrees	X Meters	Y Meters
0.	0.	.000	-5.486
0.	15.	2.186	-5.283
0.	30.	4.297	-4.681
0.	45.	6.261	-3.701
0.	60.	8.011	-2.377
0.	75.	9.488	- .752
0.	90.	10.640	1.116
15.	0.	.000	-3.470
15.	15.	1.815	-3.302
15.	30.	3.367	-2.802
15.	45.	5.198	-1.988
15.	60.	6.651	- .888
15.	75.	7.877	.460
15.	90.	8.834	2.011
30.	0.	.000	-1.683
30.	15.	1.485	-1.545
30.	30.	2.921	-1.136
30.	45.	4.256	- .470
30.	60.	5.446	.431
30.	75.	6.449	1.535
30.	90.	7.232	2.805
45.	0.	.000	.000
45.	15.	1.176	.109
45.	30.	2.312	.433
45.	45.	3.369	.960
45.	60.	4.310	1.673
45.	75.	5.105	2.547
45.	90.	5.724	3.552
60.	0.	.000	1.690
60.	15.	.865	1.770
60.	30.	1.700	2.008
60.	45.	2.477	2.396
60.	60.	3.170	2.920
60.	75.	3.754	3.563
60.	90.	4.210	4.303

Parallel of tangency: $\phi_0 = 45^0$

Central meridian: $\lambda_0 = 0^0$

parallel based on an origin of $\phi_0 = 45°$, and $\lambda_0 = 0°$. The corresponding grid is in Fig. 5.3.1.

The Lambert conformal projection with one standard parallel may be converted to a form for the transformation from the conformal sphere by letting e = 0 in Eq. (5.3.14) and (5.3.21).

$$\rho_0 = a \cot \phi_0 \qquad\qquad\qquad (5.3.23)$$

$$\rho = \rho_0 \left\{ \frac{\tan(\frac{\pi}{4} - \frac{\phi}{2})}{\tan(\frac{\pi}{4} - \frac{\phi_0}{2})} \right\}^{\sin \phi_0} . \qquad (5.3.24)$$

Equations (5.3.23) and (5.3.24) are then used in conjunction with Eq. (5.3.1) to produce the plotting equations.

The next step is to consider the Lambert conformal projection with two standard parallels. This projection has had considerable utility as aircraft navigation charts, and has been used for star charts by the U.S. Air Force. Again, the meridians are straight lines radiating from the apex of the cone, and the parallels of latitude are concentric circles.

Let the two standard parallels be chosen as ϕ_1 and ϕ_2, where $\phi_2 > \phi_1$. Then, from Eqs. (5.3.16) and (5.3.19)

$$m = \rho_1 \frac{\sin \phi_0}{R_{p1} \cos \phi_1}$$

$$= \rho_2 \frac{\sin \phi_0}{R_{p2} \cos \phi_2} \qquad (5.3.25)$$

$$\frac{\rho_1}{\rho_2} = \frac{R_{p1} \cos \phi_1}{R_{p2} \cos \phi_2} . \qquad (5.3.26)$$

From Eq. (5.3.21)

$$\frac{\rho_1}{\rho_2} = \left\{ \frac{\tan(\frac{\pi}{4} - \frac{\phi_1}{2})(\frac{1 + e \sin \phi_1}{1 - e \sin \phi_1})^{e/2}}{\tan(\frac{\pi}{4} - \frac{\phi_2}{2})(\frac{1 + e \sin \phi_2}{1 - e \sin \phi_2})^{e/2}} \right\}^{\sin \phi_0} . \qquad (5.3.27)$$

From Eqs. (5.3.26) and (5.3.27)

$$\frac{R_{p1}\cos\,\phi_1}{R_{p2}\cos\,\phi_2} = \left\{ \frac{\tan(\frac{\pi}{4} - \frac{\phi_1}{2})(\frac{1 + e\,\sin\,\phi_1}{1 - e\,\sin\,\phi_1})^{e/2}}{\tan(\frac{\pi}{4} - \frac{\phi_2}{2})(\frac{1 + e\,\sin\,\phi_2}{1 - e\,\sin\,\phi_2})^{e/2}} \right\}^{\sin\,\phi_0}$$

$$\ln\left(\frac{R_{p1}\,\cos\,\phi_1}{R_{p2}\,\cos\,\phi_2}\right) = \sin\,\phi_0\,\ln\left\{ \frac{\tan(\frac{\pi}{4} - \frac{\phi_1}{2})(\frac{1 + e\,\sin\,\phi_1}{1 - e\,\sin\,\phi_1})^{e/2}}{\tan(\frac{\pi}{4} - \frac{\phi_2}{2})(\frac{1 + e\,\sin\,\phi_2}{1 - e\,\sin\,\phi_2})^{e/2}} \right\}$$

$$\sin\,\phi_0 = \frac{\ln(\frac{R_{p1}\,\cos\,\phi_1}{R_{p2}\,\cos\,\phi_2})}{\ln\left\{ \frac{\tan(\frac{\pi}{4} - \frac{\phi_1}{2})(\frac{1 + e\,\sin\,\phi_1}{1 - e\,\sin\,\phi_1})^{e/2}}{\tan(\frac{\pi}{4} - \frac{\phi_2}{2})(\frac{1 + e\,\sin\,\phi_2}{1 - e\,\sin\,\phi_2})^{e/2}} \right\}} \cdot \qquad (5.3.28)$$

So far, the conical surface has been considered to be tangent at the central circle of parallel. In order to require secancy at ϕ_1 and ϕ_2, let m = 1 in Eq. (5.3.25)

$$\frac{\rho_1\,\sin\,\phi_0}{R_{p1}\,\cos\,\phi_1} = \frac{\rho_2\,\sin\,\phi_0}{R_{p2}\,\cos\,\phi_2} = 1$$

$$\left.\begin{array}{l} \rho_1\,\sin\,\phi_0 = R_{p1}\,\cos\,\phi_1 \\ \rho_2\,\sin\,\phi_0 = R_{p2}\,\cos\,\phi_2 \end{array}\right\} \cdot \qquad (5.3.29)$$

In Eq. (5.3.29), $\sin\,\phi_0$, as defined in Eq. (5.3.28), applies. From Eq. (5.3.21)

$$\rho_1 = \rho_0 \left\{ \frac{\tan(\frac{\pi}{4} - \frac{\phi_1}{2})(\frac{1 + e \sin \phi_1}{1 - e \sin \phi_1})^{e/2}}{\tan(\frac{\pi}{4} - \frac{\phi_0}{2})(\frac{1 + e \sin \phi_0}{1 - e \sin \phi_0})^{e/2}} \right\}^{\sin \phi_0} \cdot \qquad (5.3.30)$$

Substitute Eq. (5.3.30) into Eq. (5.3.29).

$$\rho_0 \left\{ \frac{\tan(\frac{\pi}{4} - \frac{\phi_1}{2})(\frac{1 + e \sin \phi_1}{1 - e \sin \phi_1})^{e/2}}{\tan(\frac{\pi}{4} - \frac{\phi_0}{2})(\frac{1 + e \sin \phi_0}{1 - e \sin \phi_0})^{e/2}} \right\}^{\sin \phi_0} = R_{p1} \cos \phi_1 .$$

Let an auxiliary variable be defined such that

$$\psi = \frac{\rho_0}{\tan(\frac{\pi}{4} - \frac{\phi_0}{2})(\frac{1 + e \sin \phi_0}{1 - e \sin \phi_0})^{e/2}}$$

$$= \frac{R_{p1} \cos \phi_1}{\sin \phi_0 \tan(\frac{\pi}{4} - \frac{\phi_1}{2})(\frac{1 + e \sin \phi_1}{1 - e \sin \phi_1})^{e/2}} . \qquad (5.3.31)$$

In a similar manner

$$\psi = \frac{R_{p2} \cos \phi_2}{\sin \phi_0 \tan(\frac{\pi}{4} - \frac{\phi_2}{2})(\frac{1 + e \sin \phi_2}{1 - e \sin \phi_2})^{e/2}} .$$

The polar equations become

$$\left. \begin{array}{l} \theta = \lambda \sin \phi_0 \\[2ex] \rho = \psi \{\tan(\frac{\pi}{4} - \frac{\phi}{2})(\frac{1 + e \sin \phi}{1 - e \sin \phi})^{e/2}\}^{\sin \phi_0} \end{array} \right\} \qquad (5.3.32)$$

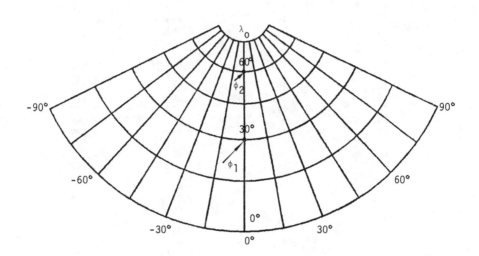

Figure 5.3.2 Lambert Conformal Projection, Two Standard Parallels

Figure 5.3.2 displays a Lambert conformal projection with two standard parallels developed by using Eq. (5.3.32) in conjunction with Eq. (5.3.1). Table 5.3.2 gives the plotting coordinates for the two standard parallel case. Notice that the meridians are straight lines, and the parallels are concentric circles. The area between the standard parallels is smaller than on the spheroid, and the area beyond the standard parallels is larger.

Equations (5.3.28), (5.3.31) and (5.3.32) can be converted to the transformation from the conformal sphere by setting e = 0.

Table 5.3.2
Lambert Conformal Projection, Two Standard Parallels

Latitude Degrees	Longitude Degrees	X Meters	Y Meters
15.	0.	.000	-2.525
15.	15.	1.909	-2.346
15.	30.	3.752	-1.814
15.	45.	5.463	- .948
15.	60.	6.983	.221
30.	0.	.000	.000
30.	15.	1.439	.135
30.	30.	2.827	.536
30.	45.	4.117	1.189
30.	60.	5.262	2.070
45.	0.	.000	2.175
45.	15.	1.034	2.272
45.	30.	2.031	2.560
45.	45.	2.953	3.029
45.	60.	3.781	3.662
60.	0.	.000	4.131
60.	15.	.669	4.194
60.	30.	1.315	4.380
60.	45.	1.915	4.684
60.	60.	2.448	5.094
75.	0.	.000	5.958
75.	15.	.329	5.989
75.	30.	.647	6.080
75.	45.	.942	6.230
75.	60.	1.204	6.431

Circles of secancy: $\phi_1 = 30^0$, $\phi_2 = 60^0$

Central meridian: $\lambda_0 = 0^0$

$$\sin \phi_0 = \frac{\ln \left(\frac{\cos \phi_1}{\cos \phi_2}\right)}{\left[\ln \frac{\left(\tan\frac{\pi}{4} - \frac{\phi_1}{2}\right)}{\left(\tan\frac{\pi}{4} - \frac{\phi_2}{2}\right)}\right]} \qquad (5.3.33)$$

$$\psi = \frac{a \cos \phi_1}{\sin \phi_0 \left[\tan\left(\frac{\pi}{4} - \frac{\phi_1}{2}\right)\right]} = \frac{a \cos \phi_2}{\sin \phi_0 \left[\tan\left(\frac{\pi}{4} - \frac{\phi_2}{2}\right)\right]} \qquad (5.3.34)$$

$$\rho = \psi \, \tan(\frac{\pi}{4} - \frac{\phi}{2}).$$ (5.3.35)

Equations (5.3.33), (5.3.34), and (5.3.35) are used with Eq. (5.3.1) to obtain the plotting equations.

Inspection of the plotting equations for the one and two standard parallel cases of the Lambert Conformal projection indicates that the direct transformation from geographic to Cartesian coordinates is relatively complex. The same is true of the inverse transformation. Consideration of an inverse transformation is deferred until Section 5.5. In that section, an approximate method of inverse transformation is introduced for the two standard parallel case. This approximation gives sufficient accuracy for practical computation.

5.4 Stereographic Projections [22], [24], [25], [26]

The stereographic projections entail the transformation from the spheroid to the plane. Three variations of the stereographic projection are derived. These are the polar, the oblique, and the equatorial.

The stereographic projection may be considered as a purely geometrical projection. This is illustrated best with the projection from the conformal sphere to the plane tangent at the pole. The geometry of this projection is given in Figure 5.4.1.

The plane is tangent to the sphere at the north pole, N. The rays emanate from the south pole, S. The principle of the stereographic projection requires that the projection point be diametrically across from the point of tangency. A typical ray from S to a point P on the earth is transformed to the position P' on the plane. Thus, the entire projection can be derived by elementary trigonometry. The same is true for the spheroid case, only the geometry is considerably more complicated.

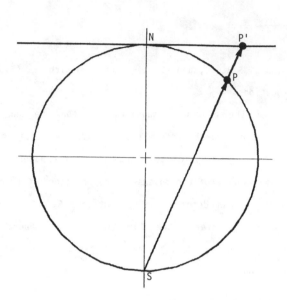

Figure 5.4.1   Geometry for the Stereographic Projection

However, the approach in this section is mathematical rather than geometrical.   This   approach   brings   out   the   quality   of   conformality immediately. We will consider the plane as one of the limiting forms of the cone, and then apply the equations already derived for the Lambert conformal projection with one standard parallel.   This is done by letting the parallel of tangency for the spheroidal case shrink to a polar point of tangency in order to derive the polar stereogrpahic projection.   To this end, let ϕ_0 = 90° in Eqs. (5.3.19), (5.3.22), (5.3.14), and (5.3.21).

$$\sin \phi_0 = 1 \qquad\qquad\qquad (5.4.1)$$

$$\theta = \lambda \qquad\qquad\qquad (5.4.2)$$

$$\rho_0 = R_{po} \cot \phi_0$$

$$= \frac{a}{\sqrt{1 - e^2}} \cot \phi_0 \tag{5.4.3}$$

and

$$\rho = \frac{a \cot \phi_0}{\sqrt{1 - e^2}} \left\{ \frac{\tan(\frac{\pi}{4} - \frac{\phi}{2})(\frac{1 + e \sin \phi}{1 - e \sin \phi})^{e/2}}{\tan(\frac{\pi}{4} - \frac{\phi_0}{2})(\frac{1 + e}{1 - e})^{e/2}} \right\}$$

$$= \tan(\frac{\pi}{4} - \frac{\phi}{2})(\frac{1 + e \sin \phi}{1 - e \sin \phi})^{e/2} \frac{a}{\sqrt{1 - e^2}} (\frac{1 - e}{1 + e})^{e/2}$$

$$\cdot \frac{\tan(\frac{\pi}{4} + \frac{\phi_0}{2})}{\tan \phi_0}. \tag{5.4.4}$$

In Eq. (5.4.4), take the limit of

$$\frac{\tan(\frac{\pi}{4} + \frac{\phi_0}{2})}{\tan \phi_0},$$

as ϕ_0 approaches 90°.

$$\lim_{\phi_0 \to 90^0} \frac{\tan(\frac{\pi}{4} + \frac{\phi_0}{2})}{\tan \phi_0} = \lim_{\phi_0 \to 90^0} \left[(\frac{1 + \tan \phi_0/2}{1 - \tan \phi_0/2})(\frac{1 - \tan^2 \phi_0/2}{2 \tan \phi_0/2}) \right]$$

$$= \lim_{\phi_0 \to 90^0} \left[\frac{(1 + \tan \phi_0/2)^2}{2 \tan \phi_0/2} \right] = 2. \tag{5.4.5}$$

Substitute Eq. (5.4.5) into Eq. (5.4.4).

$$\rho = \frac{2a}{1 - e^2} (\frac{1 - e}{1 + e})^{e/2} \tan(\frac{\pi}{4} - \frac{\phi}{2})(\frac{1 + e \sin \phi}{1 - e \sin \phi})^{e/2} \tag{5.4.6}$$

The cartesian plotting coordinates for the polar stereographic projection are

$$\left. \begin{array}{l} x = \rho \sin \theta \\ y = -\rho \cos \theta \end{array} \right\} .$$ (5.4.7)

Equations (5.4.6) are evaluated using Eqs. (5.4.3) and (5.4.6).

Equation (5.4.7) can be converted into a transformation from the conformal sphere to the plane by letting e = 0. Then

$$\rho = 2a \, \tan(\frac{\pi}{4} - \frac{\phi}{2}) .$$ (5.4.8)

Substitute Eqs. (5.4.3) and (5.4.8) into Eq. (5.4.7), and including the scale factor

$$\left. \begin{array}{l} x = 2aS \, \tan \, (\frac{\pi}{4} - \frac{\phi}{2}) \, \sin \, \Delta\lambda \\ y = -2aS \, \tan \, (\frac{\pi}{4} - \frac{\phi}{2}) \, \cos \, \Delta\lambda \end{array} \right\} .$$ (5.4.9)

Figure 5.4.2 gives an example of the polar stereographic projection. Note that the meridians are straight lines converging on the pole, and the parallels are concentric circles centered on the pole. The spacing between the parallels increases as one goes towards the equator. Table 5.4.1 gives the plotting coordinates.

The oblique case for the stereographic transformation from the conformal sphere to the plane may be obtained by applying the transformation formulas of Section 2.10. The latitude and longitude of the pole of the auxiliary coordinate system is ϕ_p and λ_p, respectively. Write Eq. (5.4.9) including the scale factor.

$$\left. \begin{array}{l} x = 2aS \, \tan \, (\frac{\pi}{4} - \frac{h}{2}) \sin \, \alpha \\ y = -2aS \, \tan \, (\frac{\pi}{4} - \frac{h}{2}) \cos \, \alpha \end{array} \right\} .$$ (5.4.10)

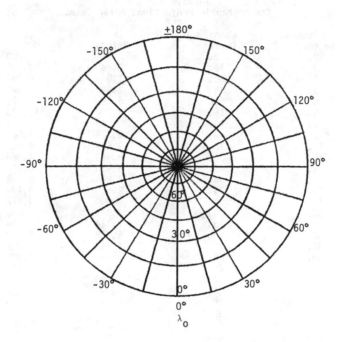

Figure 5.4.2   Stereographic Projection, Polar Case

Then, from Section 2.10.

$$\alpha = \tan\left\{\frac{\sin(\lambda - \lambda_p)}{\cos\phi_p \tan\phi - \sin\phi_p \cos(\lambda - \lambda_p)}\right\} \qquad (5.4.11)$$

$$h = \sin^{-1}\left\{\sin\phi \sin\phi_p + \cos\phi \cos\phi_p \cos(\lambda - \lambda_p)\right\}. \qquad (5.4.12)$$

Equations (5.4.10), (5.4.11), and (5.4.12) will then produce a grid such as the one in Figure 5.4.3. The only straight line in this projection is the central meridian. Table 5.4.2 is the plotting table for $\phi_p = 45°$.

The equatorial case follows when $\phi_p = 0°$ in Eqs. (5.4.11) and (5.4.12). These equations simplify to give

Table 5.4.1
Stereographic Projection, Polar Case.

Latitude Degrees	Longitude Degrees	X Meters	Y Meters
0.	0.	12.714	.000
0.	15.	12.280	3.291
0.	30.	11.010	6.357
0.	45.	8.990	8.990
0.	60.	6.357	11.010
0.	75.	3.290	12.280
0.	90.	.000	12.714
15.	0.	9.772	.000
15.	15.	9.439	2.529
15.	30.	8.463	4.886
15.	45.	6.910	6.910
15.	60.	4.886	8.463
15.	75.	2.529	9.439
15.	90.	.000	9.772
30.	0.	7.365	.000
30.	15.	7.114	1.906
30.	30.	6.378	3.682
30.	45.	5.208	5.208
30.	60.	3.682	6.378
30.	75.	1.906	7.114
30.	90.	.000	7.365
45.	0.	5.291	.000
45.	15.	5.111	1.369
45.	30.	4.582	2.646
45.	45.	3.741	3.741
45.	60.	2.645	4.582
45.	75.	1.369	5.111
45.	90.	.000	5.291
60.	0.	3.426	.000
60.	15.	3.310	.887
60.	30.	2.967	1.713
60.	45.	2.423	2.423
60.	60.	1.713	2.967
60.	75.	.887	3.310
60.	90.	.000	3.426
75.	0.	1.685	.000
75.	15.	1.627	.436
75.	30.	1.459	.842
75.	45.	1.191	1.191
75.	60.	.842	1.459
75.	75.	.436	1.627
75.	90.	.000	1.685
90.	0.	.000	.000
90.	15.	.000	.000
90.	30.	.000	.000
90.	45.	.000	.000
90.	60.	.000	.000
90.	75.	.000	.000
90.	90.	.000	.000

Point of tangency: $\phi_0 = 90^0$

Central meridian: $\lambda_0 = 0^0$

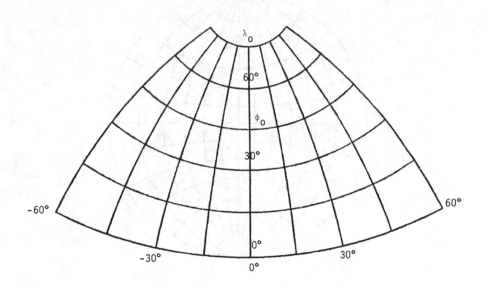

Figure 5.4.3 Stereogrpahic Projection, Oblique Case

$$\left.\begin{array}{l} \alpha = \tan^{-1} \left\{ \dfrac{\sin (\lambda - \lambda_p)}{\tan \phi} \right\} \\[2em] h = \sin^{-1} \left\{ \cos \phi \cos (\lambda - \lambda_p) \right\} \end{array}\right\} \; . \qquad (5.4.13)$$

Figure 5.4.4 shows an equatorial stereographic projection. The equator and central meridian are the only straight lines on the grid. All the other lines are arcs. The plotting coordinates are in Table 5.4.3.

The stereographic projection can also be derived for a transformation from the conformal sphere by a process similar to that introduced for the gnomonic, azimuthal equidistant and orthographic projections of Chapter 6.

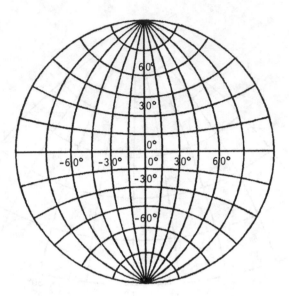

Figure 5.4.4 Stereographic Projection, Equatorial Case

Another approach to the stereographic projection can be made by using a secant rather a tangent plotting. The derivation of the plottings equation for the secant case is left as an exercise.

For the polar stereographic projection, the equations for the inverse transformation from cartesian coordinates to geographical coordinates follow from Eqs. (5.4.9) for the spherical case. Dividing the first of Eqs. (5.4.9) by the second, we obtain

$$\Delta\lambda = \tan^{-1}\left(\frac{x}{y}\right) \tag{5.4.14}$$

Summing the squares of Eqs. (5.4.9)

$$\left(\frac{x}{2aS}\right)^2 + \left(\frac{y}{2aS}\right)^2 = \tan^2\left(\frac{\pi}{4} - \frac{\phi}{2}\right) \tag{5.4.15}$$

After some manipulation of Eq. (5.4.15) the result is

Table 5.4.2
Stereographic Projection, Oblique Case

Latitude Degrees	Longitude Degrees	X Meters	Y Meters
0.	0.	.000	-5.284
0.	15.	1.962	-5.177
0.	30.	3.956	-4.845
0.	45.	6.013	-4.252
0.	60.	8.162	-3.332
0.	75.	10.416	-1.973
0.	90.	12.756	.000
15.	0.	.000	-3.418
15.	15.	1.731	-3.300
15.	30.	3.472	-2.937
15.	45.	5.230	-2.297
15.	60.	7.000	-1.326
15.	75.	8.753	.059
15.	90.	10.416	1.974
30.	0.	.000	-1.679
30.	15.	1.470	-1.561
30.	30.	2.932	-1.197
30.	45.	4.372	- .567
30.	60.	5.764	.364
30.	75.	7.057	1.646
30.	90.	8.162	3.332
45.	0.	.000	.000
45.	15.	1.177	.110
45.	30.	2.333	.442
45.	45.	3.441	1.008
45.	60.	4.464	1.822
45.	75.	5.347	2.901
45.	90.	6.013	4.252
60.	0.	.000	1.679
60.	15.	.845	1.768
60.	30.	1.662	2.036
60.	45.	2.422	2.482
60.	60.	3.087	3.106
60.	75.	3.616	3.900
60.	90.	3.956	4.845
75.	0.	.000	3.418
75.	15.	.459	3.472
75.	30.	.896	3.633
75.	45.	1.288	3.896
75.	60.	1.611	4.252
75.	75.	1.843	4.686
75.	90.	1.962	5.177
90.	0.	.000	5.284
90.	15.	.000	5.284
90.	30.	.000	5.284
90.	45.	.000	5.284
90.	60.	.000	5.284
90.	75.	.000	5.284
90.	90.	.000	5.284

Point of tangency: $\phi_0 = 45^0$

Central meridian: $\lambda_0 = 0^0$

Table 5.4.3
Stereographic Projection, Equatorial Case

Latitude Degrees	Longitude Degrees	X Meters	Y Meters
0.	0.	.000	.000
0.	15.	1.679	.000
0.	30.	3.418	.000
0.	45.	5.284	.000
0.	50.	7.365	.000
0.	75.	9.788	.000
0.	90.	12.757	.000
15.	0.	.000	1.679
15.	15.	1.850	1.708
15.	30.	3.355	1.798
15.	45.	5.177	1.962
15.	60.	7.196	2.226
15.	75.	9.522	2.641
15.	90.	12.322	3.302
30.	0.	.000	3.418
30.	15.	1.557	3.473
30.	30.	3.156	3.645
30.	45.	4.345	3.956
30.	60.	6.676	4.451
30.	75.	8.717	5.210
30.	30.	11.047	6.378
45.	0.	.000	5.284
45.	15.	1.387	5.360
45.	30.	2.797	5.594
45.	45.	4.252	6.013
45.	60.	5.771	6.664
45.	75.	7.365	7.625
45.	90.	9.020	9.020
60.	0.	.000	7.365
60.	15.	1.113	7.450
60.	30.	2.225	7.709
60.	45.	3.332	8.162
60.	60.	4.419	8.838
60.	75.	5.455	9.782
60.	90.	6.378	11.047
75.	0.	.000	8.788
75.	15.	.684	9.857
75.	30.	1.348	10.056
75.	45.	1.973	10.416
75.	60.	2.532	10.910
75.	75.	2.989	11.548
75.	90.	3.301	12.322
90.	0.	.000	12.757
90.	15.	.000	12.757
90.	30.	.000	12.757
90.	45.	.000	12.756
90.	60.	.000	12.756
90.	75.	.000	12.756
90.	90.	.000	12.756

Point of tangency: $\phi_0 = 0^0$

Central meridian: $\lambda_0 = 0^0$

Summing the squares of Eqs. (5.4.9)

$$\phi = \frac{\pi}{2} - 2\tan^{-1} \frac{x^2 + y^2}{4a^2 s^2} \qquad\qquad (5.4.16)$$

The derivations of the inverse transformations for the stereographic oblique and equatorial cases are left as exercises.

5.5 State Plane Coordinates [5], [21]

The methods used in state plane coordinate system work is an example of an approximate approach to the direct and increase transformation for two conformal projections: transverse Mercator and Lambert conformal with two standard parallels. The following paragraphs introduce state plane coordinate systems, and then give the two sets of transformation equations.

State plane coordinate systems have been developed for the entire United States to permit methods of plane surveying to be used over great distances. At the same time, a precision in the results is maintained which approaches that of geodetic surveying. Thus, a tremendous cost saving is realized, while the required level of precision is retained.

In the United States, the two map projections most used for state plane transformations are the Transverse Mercator and Lambert conformal with two standard parallels. Both of these are conformal projections, and maintain angle relation, and relative size at a point.

Geometrically, as applied in this case, the transverse Mercator is a developable cylinder secant to the earth. The axis of cylinder lies in the plane of the equator (see Figure 5.5.1a). The distance covered by each map is limited to a zone with a width of 158 miles in order to limit the amount of distortion which builds up east and west. There is no abnormal distortion north or south. Thus, the Transverse Mercator is ideally suited for states with a large north to south extent.

On the other hand, the Lambert conformal is a developable cone secant to the earth, as in Figure 5.5.1b. The axis of the cone coincides with the polar

(a) Transverse Mercator Projection:

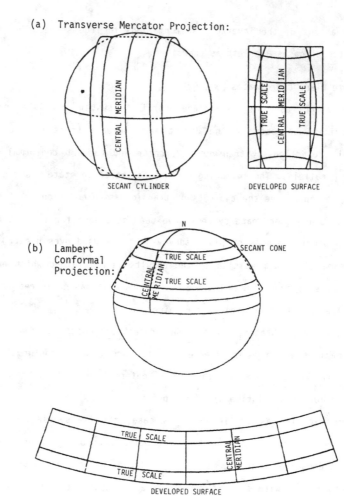

Figure 5.5.1 Mapping Surfaces for State Plane Coordinate Systems

axis of the earth. The distance covered by each map is limited to a zone with width of 158 miles, as with the transverse Mercator, in order to limit the amount of distortion which builds up north and south. There is no additional distortion east or west. Thus, the Lambert Conformal projection is ideally suited for states with a large east to west extent.

Once the decision has been made to use either the Transverse Mercator or the Lambert conformal, the state is divided into enough zones to cover the total area of the state.

For both the Lambert conformal and the Transverse Mercator projections, two transformations, the direct and the inverse are needed. In the direct transformation, the latitude and longitude of a position are known, and the cartesian state plane coordinates are required. This is a usual case for map making. In the inverse transformation, the state plane coordinates are known, and it is desired to calcualte the latitude and longitude of the position. An application of this is found in computer graphics. In this process, a map may be digitized, and then selected latitudes and longitudes may be obtained for further computation.

The National Geodetic Survey [6] has developed an approximate method for accomplishing direct and inverse transformations for both the Lambert Conformal and Transverse Mercator projections. This approximation differs slightly from the traditional methods of transformation [18], but yields an acceptable level of accuracy, especially for computer graphics applications. In addition, the National Geodetic Survey provides the necessary transformation constants.

For the Transverse Mercator, the transformation constants are defined as follows:

T_1 is the false easting, or x coordiante, of the central meridian.

T_2 is the central meridian expressed in seconds.

T_3 is the degrees and minutes portion, in minutes, of the rectifying latitude, ω_0, for ϕ_0, the latitude of the origin. (See the derivation and explanation of ω'' under L_7 of the Lambert projection.)

T_4 is the remainder of ω_0, i.e., the seconds.

T_5 is the scale along the central meridian.

$T_6 = (1/6\ R_m R_p T_5^2) \times 10^{15}$. ($R_m$ and R_p are computed for the mean latitude of the area in the zone, or projectionly Eq. (3.2.16) and Eq. (3.2.9) respectively.

For the direct transformation: geographic position to state plane coordinates, the necessary equations are:

ϕ = latitude of geographic position

λ = longitude of geographic position

$$S_1 = \frac{30.92241\ 724\ \cos\ \phi}{(1 - 0.00676\ 86580\ \sin^2\phi)^{1/2}}[T_2 - \lambda - 3.9174\ (\frac{T_2 - \lambda}{10^4})^3]$$
$$(\lambda \text{ in seconds}) \qquad (5.5.1)$$

$$S_m = S_1 + 4.0831(\frac{S_1}{10^5})^3 \qquad (5.5.2)$$

$$x = T_1 + 3.28083\ 333\ S_m T_5 + (\frac{3.28083\ 333\ S_m T_5}{10^5})^3 T_6 \qquad (5.5.3)$$

(T_1 and x will be in the millions in New Jersey; x may be in millions in certain other states.)

$$\phi_1 = \phi + \frac{25.52381}{10^{10}}\ S_m^2(1-0.00676\ 86580\ \sin^2\phi)^2\tan\ \phi \ (\phi \text{ in seconds})$$
$$(5.5.4)$$

$$\phi_2 = \phi + \frac{25.52381}{10^{10}}\ S_m^2(1 - 0.00676\ 86580\ \sin^2\ \phi_1)^2\ \tan\ \phi_1 \qquad (5.5.6)$$

$$y = 101.27940\ 65\ T_5\ \{60(\phi' - T_3) + \phi'' - T_4 - [1,052.89388\ 2$$
$$-(4.48334\ 4 - 0.02352\ 0\ \cos^2\ \phi_2)\cos^2\ \phi_2]\sin\ \phi_2\ \cos\ \phi_2\} \qquad (5.5.7)$$

ϕ' is degrees and minutes of ϕ_2 in whole minutes

ϕ'' is the remainder of ϕ_2 in seconds

For the inverse transformation: state plane coordinates to geographic
position, the equations are:

$$S_{g1} = x - T_1 - T_6 (\frac{x - T_1}{10^5})^3 \tag{5.5.8}$$

$$S_m = \frac{0.30480\ 06099}{T_5} [x - T_1 - T_6 (\frac{S_{g1}}{10^5})^3] \tag{5.5.9}$$

$$\omega' = T_3 \quad \text{(degrees and minutes of } \omega \text{ in whole minutes)} \tag{5.5.10}$$

$$\omega'' = T_4 + \frac{0.00987\ 36755\ 53}{T_5} y \quad \text{(remainder of } \omega \text{ in seconds)} \tag{5.5.11}$$

$$\omega = \omega' + \omega'' \tag{5.5.12}$$

$$(\phi')' = T_3 \quad \text{(degrees and minutes of } \phi \text{ and } \phi' \text{ in while minutes)}$$
$$\tag{5.5.13}$$

$$(\phi')'' = \omega'' + [1,047.54671\ 0 + (6.19276\ 0$$
$$+ 0.05091\ 2 \cos^2\omega)\cos^2 \omega]\sin \omega \cos \omega$$
$$\text{(remainder of } \phi' \text{ in seconds)} \tag{5.5.14}$$

$$\phi' = (\phi')' + (\phi')'' \tag{5.5.15}$$

$$(\phi)'' = (\phi')'' - 25.52381(1 - 0.00676\ 86580 \sin^2\phi')^2 (\frac{S_m}{10^5})^2 \tan \phi'$$
$$\tag{5.5.16}$$

$$\phi = (\phi')' + (\phi)'' \tag{5.5.17}$$

$$S_a = S_m - 4.0831 (\frac{S_m}{10^5})^3 \tag{5.5.18}$$

$$S_1 = S_m - 4.0831 (\frac{S_a}{10^5})^3 \tag{5.5.19}$$

$$\Delta\lambda_1 = \frac{S_1(1 - 0.00676\ 86580 \sin^2 \phi)^{1/2}}{30.92241\ 724 \cos \phi} \tag{5.5.20}$$

$$\Delta\lambda_a = \Delta\lambda_1 + 3.9174 (\frac{\Delta\lambda_1}{10^4})^3 \tag{5.5.21}$$

$$\lambda'' = T_2 - \Delta\lambda_1 - 3.9174 (\frac{\Delta\lambda_a}{10^4})^3 \tag{5.5.22}$$

A similar set of transformation constants for the Lambert conformal
projection are defined as follows:

L_1 is the false easting, or x coordinate, of the central meridian.

L_2 is the central meridian expressed in seconds.

L_3 is the map radius of the central parallel (ϕ_0).

L_4 is the map radius of the lowest parallel of the projection table
   plus the y value on the central meridian at this parallel. This y
   value is zero in most, but not all, cases.

L_5 is the scale (m) of the projection along the central parallel (ϕ_0).

L_6 is the ℓ computed from the basic equations for the Lambert
   projection with two standard parallels.

L_7 is the degrees and minutes portion, in minutes, of the rectifying
   latitude for ϕ_0, where ϕ_0 = arc sin ℓ.

L_8 is the remainder of ω_0, i.e., the seconds.

$L_9 = (1/6R_m R_p) \times 10^{16}$.

$L_{10} = [\tan \phi_0/24(R_m R_p)^{3/2}] \times 10^{24}$

$L_{11} = [(5 + 3 \tan^2 \phi_0)/120R_m R_p^3] \times 10^{32}$.

where R_m and R_p are given by Eqs. (3.2.16) and (3.2.9) respectively,
evaluated at ϕ_0.

The equations for the direct transformation: geographic position to state
plane coordinates, are given below.

ϕ = latitude of geographic position

λ = longitude of geographic position

$s = 101.27940\ 65\{60(L_7 - \phi') + L_8 - \phi'' + [1,052.89388\ 2$ (5.5.23)
$- (4.48334\ 4 - 0.02352\ 0 \cos^2 \phi) \cos^2 \phi]\sin \phi \cos \phi\}$

(ϕ' is the degrees and minutes of ϕ expressed in whole minutes)

(ϕ'' is the remainder of ϕ in seconds)

$$R = L_3 + sL_5 \{1 + (\frac{s}{10^8})^2[L_9 - (\frac{s}{10^8})L_{10} + (\frac{s}{10^8})^2 L_{11}]\}$$ (5.5.24)

$$\theta = L_6(L_2 - \lambda) \tag{5.5.25}$$

(θ and λ are in seconds)

$$x = L_4 + R \sin \theta \tag{5.5.26}$$

$$y = L_4 - R + 2R \sin^2 \frac{\theta}{2} \tag{5.5.27}$$

For the inverse transformation: state plane coordinates to geographic position, the equations are as follows:

$$\theta = \tan^{-1}(\frac{x - L_1}{L_4 - y}) \tag{5.5.28}$$

$$\lambda = L_2 - \frac{\theta}{L_6} \quad (\theta \text{ and } \lambda \text{ are in seconds}) \tag{5.5.29}$$

$$R = \frac{L_4 - y}{\cos \theta} \tag{5.5.30}$$

$$s_1 = \frac{L_4 - L_3 - y + 2R \sin^2 \frac{\theta}{2}}{L_5} \tag{5.5.31}$$

$$s_2 = \frac{s_1}{1 + (\frac{s_1}{10^8})^2 L_9 - (\frac{s_1}{10^8})^3 L_{10} + (\frac{s_2}{10^8})^4 L_{11}} \tag{5.5.32}$$

$$s_3 = \frac{s_1}{1 + (\frac{s_2}{10^8})^2 L_9 - (\frac{s_2}{10^8})^2 L_{10} + (\frac{s_2}{10^8})^4 L_{11}} \tag{5.5.33}$$

$$s = \frac{s_1}{1 + (\frac{s_3}{10^8})^2 L_9 - (\frac{s_3}{10^8})^3 L_{10} + (\frac{s_3}{10^8})^4 L_{11}} \tag{5.5.34}$$

$\omega' = L_7 - 600$ (degrees and minutes of ω in whole minutes) (5.5.35)

$\omega'' = 36,000 + L_8 - 0.00987\ 36755\ 53\ s$ (remainder of ω in seconds)

$$\tag{5.5.36}$$

$$\omega = \omega' + \omega'' \tag{5.5.37}$$

$\phi' = L_7 - 600$ (degrees and minutes of ϕ in while minutes) (5.5.38)

$\phi'' = \omega'' + [1,047.54671\ 0 + (6.19276\ 0$

$\qquad + 0.05091\ 2 \cos^2\omega)\cos^2\omega]\sin \omega \cos \omega$

(remainder of ϕ in seconds)   (5.5.39)

$\phi = \phi' + \phi''$ (5.5.40)

The transformation constants for both the Lambert conformal and the trasnverse Mercator are available in Ref. [6] for the entire United States.

In the form given above, these equations are extremely convenient for computer computation. A FORTRAN 4 program implementing these equations is given in Reference [21]. Another approach is given in Reference [18]. However, this second method is more applicable to desk calculator computations where only a few transformations are required.

PROBLEMS

5.1 Given a spheroidal earth of the WGS-72 model with $\phi_0 = 30°N$. Find the radius of the conformal sphere.

5.2 Let $\lambda_0 = 15°W$. What are the conformal latitude and longitude for geodetic $\lambda = 15°E$, $\phi = 40°N$, using the data of Problem 5.1?

5.3 Using a spherical earth model, and approximating the earth radius as 6,378,000 meters find the Mercator, cartesian coordinates for $\phi = 30°$, $\lambda = 15$, when $\lambda_0 = 0°$. S:(1 in $\equiv$ 1,000,000 meters).

5.4 For the scale factor and earth radius of problem 5.3, find the Mercator inverse transformation using the spherical model for x = 8", y = 5". Let $\lambda_0 = 0°$.

5.5 Consider the WGS-72 spheroid, $\lambda_0 = 0°$. Using the direct Mercator transformation from the spheroid to the map, find the cartesian plotting coordinates when $\phi = 30°$, $\lambda = 30°$. Let S(1 in $\equiv$ 1,000,000).

5.6 For a transverse Mercator projection $\lambda_0 = 0°$. Let the radius of the conformal sphere be 6,378,000 meters. Find the cartesian plotting coordinates when $\phi = 20°$, $\lambda = 10°$. Use S:(1" $\equiv$ 500,000 meters).

5.7 For an oblique Mercator projection, $\phi_0 = 45°$, $\lambda_0 = 0°$. The radius of the conformal sphere is 6,378,000 meters, and S(1" $\equiv$ 500,000). Find the cartesian plotting coordinates if $\phi = 22.5°$, $\lambda = 45°$.

5.8 Consider a Lambert conformal projection of one standard parallel tangent at $\phi_0 = 45°$. Use the WGS-72 model of the spheroid. Use S(1" $\equiv$ 250,000 meters). Let $\lambda_0 = 10°$. Find the cartesian plotting coordinates for $\lambda = 30°$, $\phi = 50°$.

5.9 Repeat Problem 5.8 using a conformal spherical model of the earth.

5.10 Consider a Lambert conformal projection of two standard parallels, secant at $\phi_1 = 30°$ and $\phi_2 = 60°$. Use the WGS-72 model of the spheroid, and S(1" $\equiv$ 250,000 meters). Let $\lambda_0 = 10°$. Find the cartesian plotting coordinates for $\lambda = 30°$, $\phi = 50°$.

5.11 Repeat problem 5.10 using a conformal spherical model of the earth.

5.12 Given the polar stereographic projection and the WGS-72 model of the spheroid $\lambda_0 = 0°$, and $S(1" \equiv 250,000$ meters). Find the cartesian plotting coordinates for $\lambda = 25°$, and $\phi = 80°$.

5.13 Repeat problem 5.12 using a conformal spherical model of the earth.

5.14 Using the data of problem 5.13, find the inverse polar stereograph transformation when $x = 4"$ and $y = 5"$.

5.15 Derive the direct plotting equations for a polar stereographic projection secant at $\phi_0 = 80°$. Use spherical model of the earth.

5.16 Consider an oblique stereographic projection with $\phi_0 = 45°$ and $\lambda_0 = 0°$. Use the radius of the spherical earth as 6,378,000 meters and $S(1" \equiv 250,000$ meters). Find the cartesian plotting coordinates when $\phi = 60°$, and $\lambda = 15°$.

5.17 Consider an equatorial stereographic projection with $\lambda_0 = 0°$. Use the radius of the spherical earth as 6,378,000 meters, and $S(1" \equiv 250,000$ meters). Find the cartesian plotting coordinates when $\phi = 30°$, and $\lambda = 15°$.

5.18 Given the transformation constants for a transverse Mercator state plane coordinate system below.

$T_1 = 500.000.000$ $T_4 = 18.35156$

$T_2 = 416.700.000$ $T_5 = .99993\ 33333$

$T_3 = 2.491$ $T_6 = .38062\ 27$

Find the cartesian coordinates when $\phi = 48° \ 07'50.94"$ and $\lambda = 116° \ 22' \ 02.59"$

5.19 Given the data of problem 5.18. Find ϕ and λ when $x = 349,231.30$ ft and $y = 2,357,247.28$ ft.

5.20 Given the transformation constants for a Lambert conformal state plane system below.

$L_1 = 3,000,000.000$ $L_7 = 3.161$

$L_2 = 633,600.00000$ $L_8 = 47.87068$

$L_3 = 15,893,950.36$ $L_9 = 3.79919$

$L_4 = 16,564,628.77$ $L_{10} = 5.91550$

$L_5 = .99984\ 80641$ $L_{11} = 5.91550$

$L_6 = .79692\ 23940$

Find the cartesian coordinates when $\phi = 54° \ 27' \ 30.00"$ and $\lambda = 164.02' \ 30.00"$ w.

5.21 Given the data of Problem 5.20.   Find ϕ and λ when x = 5,533,424.39 ft,
     and y = 1,473,805.13 ft.

6

Conventional Projections

Conventional projections are those which are neither equal area nor conformal. As was mentioned in Chapter 1, this is not a derogatory term. The conventional projections were produced in order to preserve some special quality which is more important to a particular cartographer than the qualities of equal area or conformality, or to present a projection which is either mathematically or graphically simple. Since the category of conventional is a catch-all, it is to be expected that there is a wide variety in this class of projections. This is true. Some of these are really of historical interest. Some are simply convenient. Others have proved to be cartographic work-horses.

The most useful of the conventional projections are the gnomonic, the azimuthal equidistant, and the polyconic, both regular and transverse. The simple geometrical projections of the conical and cylindrical, as well as the orthographic projection of the engineer, are examples of strictly geometrical approaches to the problem. Of mainly historical interest, are the Van der Grinten, the Plate Carreé, the Carte Parallelogrammatique, the Gall, the Murdoch, the Cassini, and the stereographic variations such as the Clarke, the James, and the La Hire. Finally, there are such mathematical endeavors as the globular. Clearly, the conventional projections provide maps ranging from the most utilitarian to the unique.

6.1 Gnomonic Projection [8], [22]

The gnomonic projection requries that the transformation of positions on the surface of the earth onto a plane be based upon a projection point at the center of the earth. The name comes from a gnome's-eye view of the world. The gnomonic projection can be a purely geometrical construction. However, we use spherical trigonometry to obtain the oblique gnomonic projection. Then, the two limiting cases of the projection, the polar and the equatorial, are obtained by particularizing the oblique case.

Figure 6.1.1 portrays the geometry required for deriving the oblique gnomonic projection. Let a plane be tangent to the sphere at point 0, whose coordinates are (ϕ_0, λ_0). The cartesian axes on the plane are such that x is east, and y is north. Let an arbitrary point P, with coordinates (ϕ, λ) be projected onto the plane to become P', with mapping coordinates (x,y).

Define an auxiliary angle, ψ, between the radius vectors CO and CP. From the figure

$$OP' = a \tan \psi \qquad\qquad\qquad (6.1.1)$$

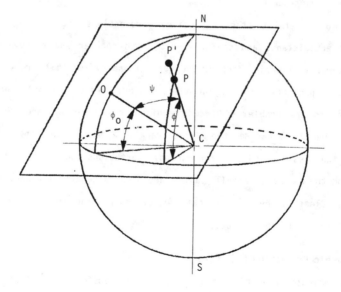

Figure 6.1.1 Geometry of the Oblique Gnomonic Projection

On the mapping plane define the second auxiliary angle, θ, which orients OP' with respect to the x-axis. This leads to

$$x = OP' \cos \theta \qquad (6.1.2)$$

Substitute Eq. (6.1.1) into Eq. (6.1.2).

$$x = a \tan \phi \cos \theta$$

$$= \frac{a \sin \phi}{\cos \phi} \cos \theta \qquad (6.1.3)$$

Also, we have

$$y = OP' \sin \theta \qquad (6.1.4)$$

Substitute Eq. (6.1.1) into Eq. (6.1.4) to obtain

$$y = a \tan \phi \sin \theta$$

$$= \frac{a \sin \phi}{\cos \phi} \sin \theta \qquad (6.1.5)$$

It is now necessary to find ψ and θ in terms of ϕ, ϕ_0, λ, and λ_0. From Figure 6.1.1, by the use of the law of sines, with $\Delta\lambda = \lambda - \lambda_0$

$$\frac{\sin \Delta\lambda}{\sin \psi} = \frac{\sin(90^0 - \theta)}{\sin(90^0 - \phi)}$$

$$= \frac{\cos \theta}{\cos \phi}$$

$$\sin \psi \cos \theta = \sin \Delta\lambda \cos \phi \qquad (6.1.6)$$

Apply the law of cosines to obtain the following.

$$\cos \psi = \cos(90^0 - \phi_0)\cos(90^0 - \phi)$$
$$+ \sin(90^0 - \phi_0)\sin(90^0 - \phi)\cos \Delta\lambda$$
$$= \sin \phi_0 \sin \phi + \cos \theta_0 \cos \phi \cos \Delta\lambda \qquad (6.1.7)$$

Apply Eq. (2.10.6)

$$\sin \psi \cos(90^0 - \theta) = \sin(90^0 - \phi_0)\cos(90^0 - \phi)$$

$$- \cos(90^0 - \phi_0)\sin(90^0 - \phi)\cos \Delta\lambda$$
$$\sin \psi \sin \theta = \cos \theta_0 \sin \phi$$
$$- \sin \phi_0 \cos \phi \cos \Delta\lambda \qquad (6.1.8)$$

Substitute Eqs. (6.1.6), and (6.1.7) into (6.1.3), and Eqs. (6.1.7) and (6.1.8) into Eq. (6.1.5).

$$\left. \begin{array}{l} x = \dfrac{aS \cos \phi \sin \Delta\lambda}{\sin \phi_0 \sin \phi + \cos \phi_0 \cos \phi \cos \Delta\lambda} \\[3mm] y = \dfrac{aS(\cos \phi_0 \sin \phi - \sin \phi_0 \cos \phi \cos \Delta\lambda)}{\sin \phi_0 \sin \phi + \cos \phi_0 \cos \phi \cos \Delta\lambda} \end{array} \right\} . \qquad (6.1.9)$$

Equations (6.1.9), with the inclusion of the scale factor, S, are the plotting equations for the oblique gnomonic projection. The grid resulting from a selection of ϕ_0 = 45°, and λ_0 = 0° is given as Figure 6.1.2. In this projection, all the meridians, and the equator are straight lines, since they are great circles. An arbitrary great circle distance between points A and C, on the figure, is also a straight line. The loxodrome between these same two points appears as a curved line. Compare Figure 6.1.2 to Figure 5.2.3 for the equatorial Mercator, in which the situation is reversed. Table 6.1.1 has the plotting coordinates.

To find the gnomonic polar projection, it is necessary to let ϕ_0 = 90° in Eq. (6.1.9).

$$x = - \frac{aS \cos \phi \sin \Delta\lambda}{\sin \phi}$$

$$= - aS \cot \phi \sin \Delta\lambda \qquad (6.1.10)$$

$$y = \frac{aS \cos \phi \cos \Delta\lambda}{\sin \phi}$$

$$= aS \cot \phi \cos \Delta\lambda \qquad (6.1.11)$$

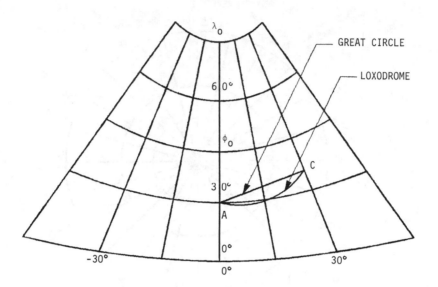

Figure 6.1.2 Gnomonic Projection, Oblique Case

A polar gnomonic grid is given in Figure 6.1.3, and based on Eqs. (6.1.10) and (6.1.11). In this case, all meridians again are straight lines. The parallels are concentric circles, whose spacing increases as the latitude decreases. Thus, the distortion becomes extreme as the equator is approached. The equator itself can never be portrayed on the gnomonic polar projection, since a ray from the center of the earth to any point on the equator will be parallel to the projection plane. Table 6.1.2 is the plotting table for the polar projection.

The equatorial gnomonic projection is obtained from Eq. (6.1.9) by setting $\phi_0 = 0^0$.

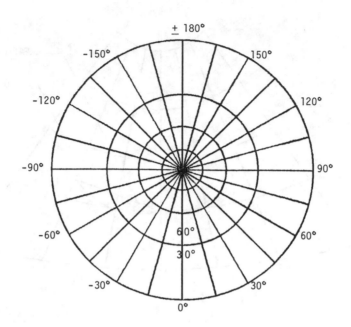

Figure 6.1.3 Gnomonic Projection, Polar Case

$$x = - \frac{aS \cos \phi \sin \Delta\lambda}{\cos \phi \cos \Delta\lambda}$$

$$= - aS \tan \Delta\lambda$$

$$y = \frac{aS \sin \phi}{\cos \phi \cos \Delta\lambda} \qquad\qquad (6.1.12)$$

Figure 6.1.4 gives the equatorial gnomonic projection based on Eq. (6.1.12).
Again, only the meridians and the equator are straight lines. The plotting
coordinates are in Table 6.1.3.

Table 6.1.1
Gnomonic Projection, Oblique Case

Latitude Degrees	Longitude Degrees	X Meters	Y Meters
15.	0.	.000	-3.682
15.	15.	1.892	-3.608
15.	30.	3.977	-3.364
15.	45.	6.541	-2.873
30.	0.	.000	-1.709
30.	15.	1.513	-1.606
30.	30.	3.125	-1.276
30.	45.	4.966	- .644
45.	0.	.000	.000
45.	15.	1.188	.111
45.	30.	2.417	.458
45.	45.	3.736	1.094
60.	0.	.000	1.709
60.	15.	.865	1.811
60.	30.	1.736	2.126
60.	45.	2.615	2.680
75.	0.	.000	3.682
75.	15.	.497	3.755
75.	30.	.981	3.976
75.	45.	1.437	4.346

Latitude of tangency: $\phi_0 = 45^0$

Central meridian: $\lambda_0 = 0^0$

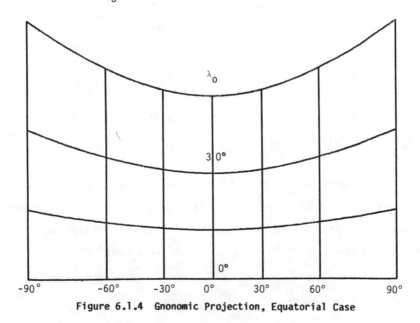

Figure 6.1.4 Gnomonic Projection, Equatorial Case

Table 6.1.2
Gnomonic Projection, Polar Case

Latitude Degrees	Longitude Degrees	X Meters	Y Meters
30.	0.	.000	-11.048
30.	15.	2.859	-10.671
30.	30.	5.524	- 9.567
30.	45.	7.812	- 7.812
30.	60.	9.567	- 5.524
30.	75.	10.671	- 2.859
30.	90.	11.047	.000
45.	0.	.000	- 6.378
45.	15.	1.651	- 6.161
45.	30.	3.189	- 5.524
45.	45.	4.510	- 4.510
45.	60.	5.524	- 3.189
45.	75.	6.161	- 1.651
45.	90.	6.378	.000
60.	0.	.000	- 3.682
60.	15.	.953	- 3.557
60.	30.	1.841	- 3.189
60.	45.	2.604	- 2.604
60.	60.	3.189	- 1.841
60.	75.	3.557	- .953
60.	90.	3.682	.000
75.	0.	.000	- 1.709
75.	15.	.442	- 1.651
75.	30.	.854	- 1.480
75.	45.	1.208	- 1.208
75.	60.	1.580	- .855
75.	75.	1.651	- .442
75.	90.	1.709	.000
90.	0.	.000	.000
90.	15.	.000	.000
90.	30.	.000	.000
90.	45.	.000	.000
90.	60.	.000	.000
90.	75.	.000	.000
90.	90.	.000	.000

Central meridian: $\lambda_0 = 0^0$

The inverse polar and equatorial gnomonic transformations follow in a straightforward manner from their direct counterparts. For the inverse transformation from cartesian to geographic coordinates for the gnomonic polar projection, inversion of Eqs. (6.1.10) and (6.1.11) gives

$$\Delta\lambda = \tan^{-1} \left(-\frac{x}{y}\right)$$
$$\phi = \tan^{-1} \frac{(aS)^2}{x^2 + y^2}$$

(6.1.13)

The corresponding inverse transformation for the equatorial gnomonic projection is obtained by the inversion of Eqs. (6.1.12)

$$\Delta\lambda = \tan^{-1} \left(\frac{x}{aS}\right)$$

$$\phi = \sin^{-1} \left(\frac{y \sin \Delta\lambda}{-x}\right) \qquad\qquad (6.1.14)$$

The inverse transformation for the oblique case is much more complicated, and is left as an exercise.

Table 6.1.3
Gnomonic Projection, Equatorial Case

Latitude Degrees	Longitude Degrees	X Meters	Y Meters
0.	0.	.000	.000
0.	15.	1.709	.000
0.	30.	3.682	.000
0.	45.	6.378	.000
15.	0.	.000	1.709
15.	15.	1.709	1.769
15.	30.	3.682	1.973
15.	45.	6.378	2.417
30.	0.	.000	3.682
30.	15.	1.709	3.812
30.	30.	3.682	4.252
30.	45.	6.378	5.208
45.	0.	.000	6.378
45.	15.	1.709	6.603
45.	30.	3.682	7.365
45.	45.	6.378	9.020

Central meridian: $\lambda_0 = 0^0$

6.2 Azimuthal Equidistant Projection [8], [22]

The azimuthal equidistant projection (also called Postel's projection) is another projection directly from the earth onto a plane. The projection law in this case is that the distance and bearing from the origin of the plotting surface to any other point must be true. Thus, all great circles through the origin are lines of true length.

Figure 6.2.1 shows the geometry for the oblique azimuthal equidistant projection. Again, the plane is tangent to the spherical earth at the mapping

origin, O. The coordinates of the origin are (ϕ_0, λ_0). An arbitrary point P
on the sphere has coordinates (ϕ, λ). Again, the auxiliary angle between the
radius vectors CO and CP is ψ. By the definition of the law of the
equidistant transformation

$$OP' = a\psi \qquad (6.2.1)$$

$$\left. \begin{array}{l} x = OP' \cos \theta \\ y = OP' \sin \theta \end{array} \right\} . \qquad (6.2.2)$$

Substitute Eq. (6.2.1) into Eq. (6.2.2).

$$\left. \begin{array}{l} x = a\psi \cos \theta \\ \\ y = a\psi \sin \theta \end{array} \right\} . \qquad (6.2.3)$$

The value of ψ is found from Eq. (6.1.7).

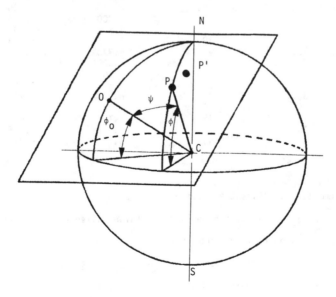

Figure 6.2.1 Geometry for the Oblique Azimuthal Equidistant Case

$$\psi = \cos^{-1}(\sin \phi_0 \sin \phi + \cos \phi_0 \cos \phi \cos\Delta\lambda) \qquad (6.2.4)$$

where $\Delta\lambda = \lambda - \lambda_0$. Since ψ is restricted to the range from $0°$ to $180°$, ψ is uniquely defined. Then, $\sin \psi$ is available immediately. Equations (6.1.6) and (6.1.7) can then be used to obtain θ.

$$\cos \theta = \frac{\sin \Delta\lambda \cos \phi}{\sin \psi} \qquad (6.2.5)$$

$$\sin \theta = \frac{\cos \phi_0 \sin \phi - \sin \phi_0 \cos \phi \cos \Delta\lambda}{\sin \psi} \qquad (6.2.6)$$

Equations (6.2.4), (6.2.5), (6.2.6) and (6.2.3), with the introduction of the scale factor, S, are used to produce an oblique azimuthal equidistant grid. Such a grid appears in Figure (6.2.2). Only the central meridian is a straight line. All other meridians, the parallels, and the equator appear as curves of varying degrees of curvature. However, any straight line ruled on the map, from the origin to any arbitrary point will be true length, and true azimuth. The azimuthal equidistant projection has seen much modern use as rocket and missile firing charts, and air route planning charts. Table 6.2.1 gives the plotting coordinates.

In order to obtain the polar azimuthal equidistant projection, let $\phi_0 = 90°$ in Eqs. (6.2.4), (6.2.5), and (6.2.6). We then have

$$\psi = \cos^{-1} (\sin \phi) \qquad (6.2.7)$$

$$\cos \psi = \sin \phi$$

$$= \cos(\pi/2 - \phi)$$

$$\psi = \frac{\pi}{2} - \phi \qquad (6.2.8)$$

Substitute Eq. (6.2.8) into eqs. (6.2.5) and (6.2.6) to obtain

$$\cos \theta = \frac{\sin \Delta\lambda \cos \phi}{\sin(\pi/2 - \phi)}$$

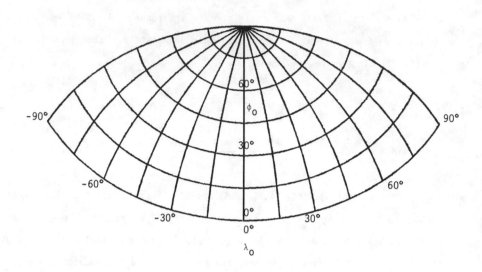

Figure 6.2.2   Azimuthal Equidistant Projection, Oblique Case

Table 6.2.1
Azimuthal Equidistant Projection, Oblique Case

Latitude Degrees	Longitude Degrees	X Meters	Y Meters
0.	0.	.000	-5.009
0.	30.	3.678	-4.504
0.	60.	7.142	-2.916
0.	90.	10.019	.000
30.	0.	.000	-1.670
30.	30.	2.874	-1.173
30.	60.	5.413	.342
30.	90.	7.142	2.916
60.	0.	.000	1.670
60.	30.	1.639	2.008
60.	60.	2.974	2.992
60.	90.	3.678	4.504
90.	0.	.000	5.009
90.	30.	.000	5.009
90.	60.	.000	5.009
90.	90.	.000	5.009

Latitude of tangency: $\phi_0 = 45^0$

Central meridian: $\lambda_0 = 0^0$

$$= \sin \Delta\lambda \tag{6.2.9}$$

$$\sin \theta = - \frac{\cos \phi \cos \Delta\lambda}{\sin(\pi/2 - \phi)}$$

$$= - \cos \Delta\lambda \tag{6.2.10}$$

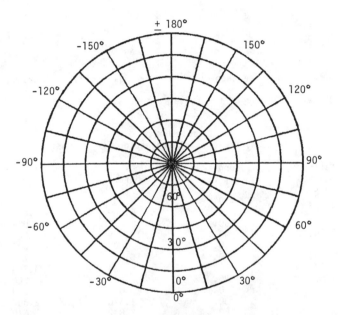

Figure 6.2.3 Azimuthal Equidistant Projection, Polar Case

Equations (6.2.3), (6.2.7), (6.2.9) and (6.2.10), with the inclusion of the scale factor, S, give the plotting equations used to develop a grid such as Figure 6.2.3. In this figure, all the meridians are straight lines of true length, and the parallels are concentric circles, equally spaced. The coordinates for the polar case are in Table 6.2.2.

The polar azimuthgal equidistant projection is the simple polar grid used in plotting engineering data.   It serves as a reasonable diagrammatic representation if no greater sophistication is required.

The equatorial azimuthal equidistant projection is obtained by substituting $\phi_0 = 0°$ into Eqs. (6.2.4) and (6.2.6).

$$\psi = \cos^{-1}[\cos \phi \cos \Delta\lambda] \qquad (6.2.11)$$

$$\sin \theta = \frac{\sin \phi}{\sin \psi} \qquad (6.2.12)$$

The plotting equations are then obtained from Eqs. (6.2.3), (6.2.5), (6.2.11) and (6.2.12), with the aid of the scale factor, S.

Table 6.2.2
Azimuthal Equidistant Projection, Polar Case

Latitude Degrees	Longitude Degrees	X Meters	Y Meters
0.	0.	.000	-10.019
0.	30.	5.009	- 8.677
0.	60.	8.677	- 5.009
0.	90.	10.019	.000
30.	0.	.000	- 6.679
30.	30.	3.340	- 5.784
30.	60.	5.784	- 3.340
30.	90.	6.679	.000
60.	0.	.000	- 3.340
60.	30.	1.670	- 2.892
60.	60.	2.892	- 1.670
60.	90.	3.339	.000
90.	0.	.000	.000
90.	30.	.000	.000
90.	60.	.000	.000
90.	90.	.000	.000

Central meridian: $\lambda_0 = 0°$

6.3 Orthographic Projection [22], [24]

The orthographic projection is yet another means of portraying the sphere upon the plane by a direct transformation.   This is another projection that can be developed by a purely graphical means.   In the orthographic projeciton, the perspective point is placed at infinity.   The projection rays fall perpendicularly upon the tangent mapping plane, after intersecting the sphere.   The geometry of this projection is shown in Figure 6.3.1 for the oblique case.   Only a hemisphere or less can be portrayed on this projection.

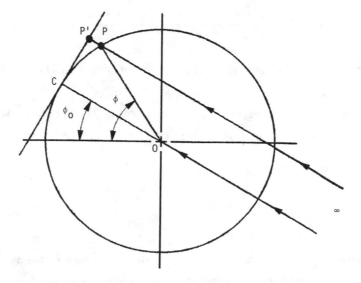

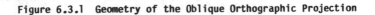

Figure 6.3.1 Geometry of the Oblique Orthographic Projection

Again, the auxiliary angle, ϕ, between CO and CP, and the auxiliary angle, θ, on the mapping plane are needed.   From the figure

$$OP' = aS \sin \phi \qquad\qquad (6.3.1)$$

$$\left.\begin{array}{l} x = OP' \cos \theta \\ y = OP' \sin \theta \end{array}\right\} \cdot \qquad\qquad (6.3.2)$$

Substitute Eq. (6.3.1) into Eq. (6.3.2)

$$x = aS \sin \psi \cos \theta \left.\begin{array}{c}\\\end{array}\right\}$$
$$y = aS \sin \psi \sin \theta \left.\begin{array}{c}\\\end{array}\right. \qquad (6.3.3)$$

where S is the scale factor.

Equation (6.1.7) again gives

$$\psi = \cos^{-1} \{\sin \phi_0 \sin \phi + \cos \phi_0 \cos \phi \cos \Delta\lambda\} \qquad (6.3.4)$$

with $\sin \psi$ readily available, since $0 < \psi < 90$. From Eqs. (6.1.6) and (6.1.8)

$$\theta = \tan^{-1} \left(\frac{\cos \phi_0 \sin \phi - \sin \phi_0 \cos \phi \cos \Delta\lambda}{\sin \Delta\lambda \cos \phi}\right) \qquad (6.3.5)$$

Equations (6.3.3), (6.3.4), and (6.3.5) give the oblique orthographic projection.

The polar orthographic projection can be obtained from the oblique projection by letting $\phi_0 = 90°$ in Eqs. (6.3.4) and (6.3.5).

$$\psi = \cos^{-1} (\sin \phi) \qquad (6.3.6)$$

$$\theta = \tan^{-1} \left[\frac{-\cos \Delta\lambda}{\sin \Delta\lambda}\right] \qquad (6.3.7)$$

Equations (6.3.3), (6.3.6), and (6.3.7) yield a grid such as the one in Figure 6.3.2. The meridians are straight lines, and the parallels are concentric circles. As the equator is approached the parallel circles are compressed together, and distortion becomes extreme. The plotting coordinates are in Table 6.3.1.

An equatorial orthographic projection is given by Figure 6.3.3. This type of projection was used to produce the first of the Lunar maps. The grid is obtained by letting $\phi_0 = 0°$ in Eqs. (6.3.4) and (6.3.5). This is another limiting case of the oblique projection.

$$\psi = \cos^{-1}[\cos \phi \cos \Delta\lambda] \qquad (6.3.8)$$

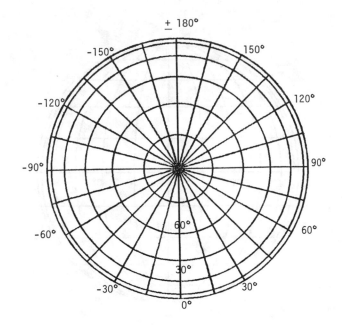

Figure 6.3.2   Orthographic Polar Projection

Table 6.3.1
Orthographic Projection, Polar Case

Latitude Degrees	Longitude Degrees	X Meters	Y Meters
0.	0.	.000	-6.378
0.	30.	3.189	-5.524
0.	60.	5.524	-3.189
0.	90.	6.378	.000
30.	0.	.000	-5.524
30.	30.	2.762	-4.784
30.	60.	4.784	-2.762
30.	90.	5.524	.000
60.	0.	.000	-3.189
60.	30.	1.595	-2.762
60.	60.	2.762	-1.595
60.	90.	3.189	.000
90.	0.	.000	.000
90.	30.	.000	.000
90.	60.	.000	.000
90.	90.	.000	.000

Central meridian: $\lambda_0 = 0^0$

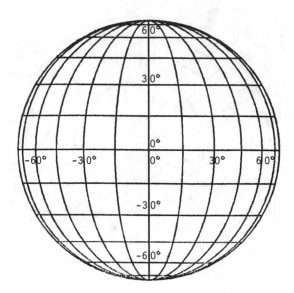

Figure 6.3.3 Orthographic Projection, Equatorial Case

Table 6.3.2
Orthographic Projection, Equatorial Case

Latitude Degrees	Longitude Degrees	X Meters	Y Meters
0.	0.	.000	.000
0.	30.	3.189	.000
0.	60.	5.524	.000
0.	90.	6.378	.000
30.	0.	.000	3.189
30.	30.	2.762	3.189
30.	60.	4.784	3.189
30.	90.	5.524	3.189
60.	0.	.000	5.524
60.	30.	1.595	5.524
60.	60.	2.762	5.524
60.	90.	3.189	5.524
90.	0.	.000	6.378
90.	30.	.000	6.378
90.	60.	.000	6.378
90.	90.	.000	6.378

Central meridian: $\lambda_0 = 0^0$

$$\theta = \tan^{-1}\left[\frac{\tan \phi}{\sin \Delta\lambda}\right] \qquad (6.3.9)$$

Equations (6.3.8) and (6.3.9) are then used with Eq. (6.3.3) to produce the required grid. In the figure, the central meridian and the equator are the only straight lines. Notice, again, from the figure that distortion becomes extreme at the margins of the map. Plotting coordinates are in Table 6.3.2.

6.4 Simple Conical Projections [8], [24]

The simple conical projections to be considered in this section are the one and two standard parallel, and the perspective cases. All of these are basically graphical projections.

The geometry for the simple conical projection, with one standard parallel, is displayed in Figure 6.4.1. The cone is tangent to the sphere at the latitude ϕ_0, with central meridian at longitude λ_0. This is displayed in section, with the cone tangent at point 0. The constant of the cone, from Section 1.12 is

$$c = \sin \phi_0 \qquad (6.4.1)$$

From the figure

$$\rho_0 = a \cot \phi_0 \qquad (6.4.2)$$

To obtain the spacing of the parallels, let the central meridian be divided truly. Thus, with the help of Eq. (6.4.2)

$$\rho = \rho_0 - a(\phi - \phi_0) \qquad (6.4.3)$$

From (6.4.1)

$$\theta = c\Delta\lambda \qquad (6.4.4)$$

Substitute Eq. (6.4.1) into Eq. (6.4.4)

$$\theta = \Delta\lambda \sin \phi_0 \qquad (6.4.5)$$

The abscissa is, from Eq. (6.4.2) and Eq. (6.4.3)

$$x = aS[\cot \phi_0 - (\phi - \phi_0)]\sin [\Delta\lambda \sin \phi_0] \qquad (6.4.6)$$

The ordinate is

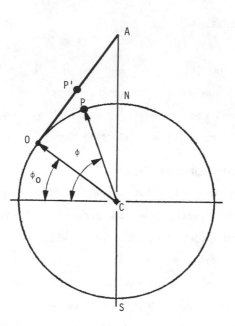

Figure 6.4.1 **Geometry of the Simple Conical Projection with one Standard Parallel**

$$y = aS\{\cot \phi_0 - [\cot \phi_0 - \phi - \phi_0]\cos [\Delta\lambda \sin \phi_0]\} \qquad (6.4.7)$$

where S is the scale factor.

The grid for this projectin is given in Figure 6.4.2 for $\phi_0 = 45^0$ and $\lambda_0 = 0°$. All of the meridians are straight lines, and the parallels are concentric circles, equally spaced. This grid has frequently been used in atlases. Table 6.4.1 gives the plotting coordinates.

The simple conical construction for the two standard parallels case follows from Figure 6.4.3. The cone is defined to have true length standard parallels at ϕ_1 and ϕ_2, with $\phi_2 > \phi_1$. From the equal spacing criterion along the central meridian we obtain the following.

$$\rho_1 - \rho_2 = a(\phi_2 - \phi_1) \tag{6.4.8}$$

From the similar triangles in Figure 6.4.3

$$\frac{\rho_1}{\rho_2} = \frac{a \cos \phi_1}{a \cos \phi_2}$$

$$= \frac{\cos \phi_1}{\cos \phi_2}$$

$$\rho_2 = \rho_1 \frac{\cos \phi_2}{\cos \phi_1} \tag{6.4.9}$$

Substitute Eq. (6.4.9) into Eq. (6.4.8).

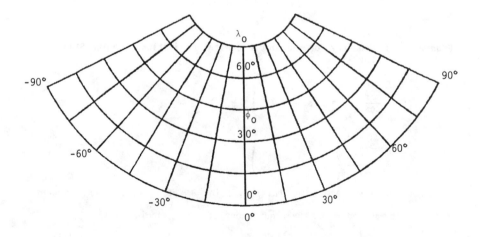

Figure 6.4.2 Simple Conical Projection, With One Standard Parallel

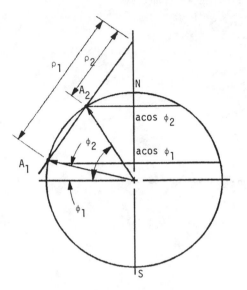

Figure 6.4.3 Geometry for the Simple Conical Projection With Two Standard Parallels

$$\rho_1\left(1 - \frac{\cos \phi_2}{\cos \phi_1}\right) = a(\phi_2 - \phi_1)$$

$$\rho_1 = \frac{a(\phi_2 - \phi_1)}{1 - \dfrac{\cos \phi_2}{\cos \phi_1}} \qquad\qquad (6.4.10)$$

A radius vector to an arbitrary point P' of latitude ϕ on the central meridian is applied, with the requirement of equal spacing

$$\rho = \rho_1 - a(\phi - \phi_1) \qquad\qquad (6.4.11)$$

Substitute Eq. (6.4.10) into Eq. (6.4.11).

$$\rho = a\left[\frac{\phi_2 - \phi_1}{1 - \dfrac{\cos \phi_2}{\cos \phi_1}} - (\phi - \phi_1)\right] \qquad\qquad (6.4.12)$$

Table 6.4.1
Simple Conical Projection, One Standard Parallel

Latitude Degrees	Longitude Degrees	X Meters	Y Meters
0.	0.	.000	-5.009
0.	15.	2.096	-4.815
0.	30.	4.121	-4.238
0.	45.	6.004	-3.298
0.	60.	7.683	-2.027
0.	75.	9.099	- .470
0.	90.	10.204	1.322
15.	0.	.000	-3.340
15.	15.	1.789	-3.174
15.	30.	3.516	-2.681
15.	45.	5.124	-1.879
15.	60.	6.556	- .795
15.	75.	7.764	.534
15.	90.	8.707	2.063
30.	0.	.000	-1.670
30.	15.	1.481	-1.532
30.	30.	2.912	-1.124
30.	45.	4.243	- .460
30.	60.	5.430	.438
30.	75.	6.430	1.539
30.	90.	7.211	2.805
45.	0.	.000	.000
45.	15.	1.174	.109
45.	30.	2.308	.432
45.	45.	3.363	.959
45.	60.	4.303	1.670
45.	75.	5.096	2.543
45.	90.	5.715	3.546
60.	0.	.000	1.670
60.	15.	.867	1.750
60.	30.	1.704	1.989
60.	45.	2.482	2.377
60.	60.	3.176	2.903
60.	75.	3.762	3.547
60.	90.	4.219	4.288

Parallel of tangency: $\phi_0 = 45^0$

Central meridian: $\lambda_0 = 0^0$

The next step is to find a constant of the cone for this configuration. From the requirement of the circle of parallel to be true length at ϕ_1.

$$2\pi a \cos \phi_1 = 2\pi c_1 \rho_1$$

$$c_1 = \frac{a \cos \phi_1}{\rho_1} \qquad\qquad (6.4.13)$$

Substitute Eq. (6.4.10) into Eq. (6.4.13)

$$c_1 = \cfrac{a \cos \phi_1}{a(\phi_2 - \phi_1)} \cdot \cfrac{1}{1 - \cfrac{\cos \phi_2}{\cos \phi_1}}$$

$$= \cos \phi_2 \cfrac{1 - \cfrac{\cos \phi_2}{\cos \phi_1}}{\phi_2 - \phi_1}$$

$$= \frac{\cos \phi_1 - \cos \phi_2}{\phi_2 - \phi_1} \qquad\qquad (6.4.14)$$

We now have the Eqs. (6.4.12) and (6.4.14) for a polar representation of the map point. The next step is to obtain the cartesian plotting equations. These are

$$x = aS \left\{ \cfrac{\phi_2 - \phi_1}{1 - \cfrac{\cos \phi_2}{\cos \phi_1}} - \phi + \phi_1 \right\} \sin [\Delta\lambda c_1] \qquad (6.4.15)$$

$$y = aS \left\{ \cfrac{\phi_2 - \phi_1}{1 - \cfrac{\cos \phi_2}{\cos \phi_1}} - \left\{ \cfrac{\phi_2 - \phi_1}{1 - \cfrac{\cos \phi_2}{\cos \phi_1}} - \phi + \phi_1 \right\} \cos \Delta\lambda c_1 \right\} \qquad (6.4.16)$$

Equations (6.4.14), (6.4.15), and (6.4.15) yield the grid of Figure 6.4.4. Again, the meridians are straight lines, and the parallels are equally spaced concentric circles. This projection has been used quite often for atlas maps where it is not necessary to have either conformality of equal area. The plotting coordinates are in Table 6.4.2.

The geometry for the conical perspective projection is shown in Figure 6.4.5. The cone is tangent at latitude ϕ_0, and the central meridian has longitude λ_0.

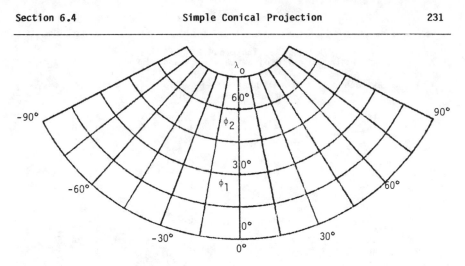

Figure 6.4.4 Simple Conical Projection With Two Standard Parallels

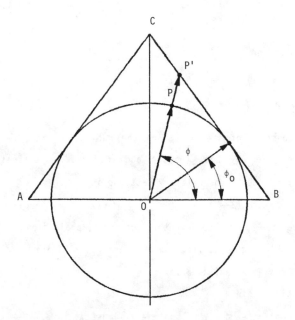

Figure 6.4.5 Geometry for the Conical Perspective Projection

Table 6.4.2
Simple Conical Projection, Two Standard Parallels

Latitude Degrees	Longitude Degrees	X Meters	Y Meters
15.	0.	.000	-1.670
15.	15.	1.742	-1.510
15.	30.	3.426	-1.036
15.	45.	4.995	- .263
15.	60.	6.398	.782
30.	0.	.000	.000
30.	15.	1.438	.132
30.	30.	2.828	.523
30.	45.	4.124	1.161
30.	60.	5.281	2.024
45.	0.	.000	1.670
45.	15.	1.134	1.774
45.	30.	2.230	2.083
45.	45.	3.252	2.586
45.	60.	4.165	3.266
60.	0.	.000	3.340
60.	15.	.830	3.416
60.	30.	1.633	3.642
60.	45.	2.381	4.010
60.	60.	3.049	4.508
75.	0.	.000	5.009
75.	15.	.526	5.058
75.	30.	1.035	5.201
75.	45.	1.509	5.435
75.	60.	1.933	5.750

Parallels of secancy: $\phi_1 = 30^0$, $\phi_2 = 60^0$

Central merdian: $\lambda_0 = 0^0$

The constant of the cone, and the radius of the parallel circle of tangency are given by Eqs. (6.4.1) and (6.4.2), respectively. From the figure, the distance to an arbitrary latitude is

$$\rho = \rho_0 - a \tan (\phi - \phi_0) \qquad (6.4.17)$$

We now have the polar coordinates for this projection. The cartesian coordinates, using Eqs. (6.4.1), (6.4.2), and (6.4.17) are

$$x = aS[\cot \phi_0 - \tan(\phi - \phi_0)]$$
$$x \sin(\Delta\lambda \sin \phi_0) \qquad (6.4.18)$$

$$y = aS\{\cot \phi_0 - [\cot \phi_0 - \tan(\phi - \phi_0)]\cos(\Delta\lambda \sin \phi_0)\} \qquad (6.4.19)$$

where S is the scale factor.

Equations (6.4.18) and (6.4.19) give the grid of Figure 6.4.6. The parallels are concentric circles, and the meridians are straight lines. The spacing of the parallels increases in either direction from the standard parallel. Thus, distortion increases significantly as one moves north or south of the circle of tangency. This projection has been used for purely illustrative atlases.

6.5 Polyconic Projection [1], [29]

The polyconic projection is a modified conical projection based on a variation of the simple conical projection. The essence of the variation is that every parallel is a standard parallel, and there are an infinity of tangent cones.

First, it is necessary to derive the polar coordinates. The central meridian is true length, and has longitude λ_0. Choose some latitude ϕ_0 as the extremity of the map, and the origin of the coordinate system. Then, the distance along the meridian for a spherical earth is

$$d = a(\phi - \phi_0) \tag{6.5.1}$$

From Eq. (1.12.1), the radius to the point of tangency for an arbitrary latitude ϕ, or the first polar coordinate, is

$$\rho = a \cot \phi \tag{6.5.2}$$

and the constant of the cone is, from Eq. (1.12.4)

$$c = \sin \phi \tag{6.5.3}$$

The second polar coordinate is, from Eq. (6.5.3)

$$\theta = \Delta\lambda \sin \phi \tag{6.5.4}$$

The Cartesian mapping equations are

$$x = \rho \sin \theta$$
$$y = d + \rho(1 - \cos \theta) \tag{6.5.5}$$

Substituting Eqs. (6.5.1), (6.5.2), and (6.5.4) into Eq. (6.5.5).

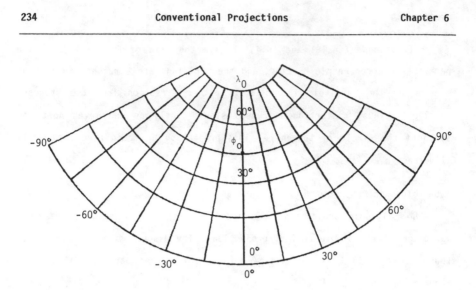

Figure 6.4.6 Perspective Conical Projection

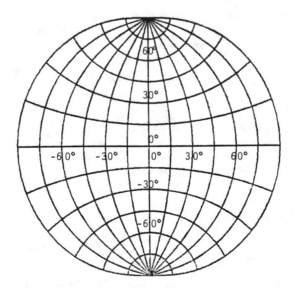

Figure 6.5.1 Polyconic Projection

$$x = aS \cot \phi \sin(\Delta\lambda \sin \phi)$$

$$\left. y = aS\{\phi - \phi_0 + \cot \phi[1 - \cos(\Delta\lambda \sin \phi)]\} \right\}.$$

$\qquad\qquad\qquad\qquad\qquad\qquad\qquad\qquad\qquad\qquad$ (6.5.6)

where S is the scale factor.

Figure 6.5.1 is the polyconic projection. The central meridian and the equator are the only straight lines. All other meridians are curves. The central meridian, the equator and all parallels are true length. Thus, the distortion occurs in angles, areas, and meridianal length for all meridians except for the central meridian. The polyconic projection has been used quite often in atlas and road maps, and its wide acceptance has justified its existence. A projection table for the polyconic projection is given in Table 6.5.1.

6.6 Simple Cylindrical Projections [24]

Two simple cylindrical projections are be considered. These are the perspective and the Miller. In both cases, the sphere is transformed to the intermediate developable surface, the cylinder.

The perspective projection is a graphical representation.

Figure 6.6.1 shows the grid of the cylindrical perspective projection. The abscissa of the plotting equations is simply

$$x = aS \Delta\lambda \qquad\qquad\qquad\qquad\qquad\qquad (6.6.1)$$

where S is the scale factor, and $\Delta\lambda = \lambda - \lambda_0$ where λ_0 is the longitude of the central meridian. The ordinate follows from consideration of the Figure

$$y = aS \tan \phi \qquad\qquad\qquad\qquad\qquad\qquad (6.6.2)$$

Then, Eqs. (6.6.1) and (6.6.2) are evaluated to obtain the grid. Distortion becomes very great as higher latitudes are reached. Thus, this projection has served in the role of an illustration.

The Miller projection calls for the equator to be 4a in length, and the meridian to be πa in length. Thus, the meridians are true, but the equator is compressed. The total area of the map is $4\pi a^2$, which is, by design, equal to the total area of the sphere. The plotting equations are simply

Table 6.5.1
Regular Polyconic Projection

Latitude Degrees	Longitude Degrees	X Meters	Y Meters
0.	0.	.000	.000
0.	30.	3.340	.000
0.	60.	6.679	.000
0.	90.	10.019	.000
0.	120.	13.359	.000
0.	150.	16.698	.000
0.	180.	20.038	.000
30.	0.	.000	3.340
30.	30.	2.859	3.716
30.	60.	5.524	4.820
30.	90.	7.812	6.575
30.	120.	9.567	8.863
30.	150.	10.671	11.528
30.	180.	11.047	14.387
60.	0.	.000	6.679
60.	30.	1.613	7.051
60.	60.	2.900	8.093
60.	90.	3.601	9.592
60.	120.	3.574	11.248
60.	150.	2.825	12.724
60.	180.	1.504	13.723
90.	0.	.000	10.019
90.	30.	.000	10.019
90.	60.	.000	10.019
90.	90.	.000	10.019
90.	120.	.000	10.019
90.	150.	.000	10.019
90.	180.	.000	10.019

Map Origin: $\phi_0 = 0^0$

Central Meridian: $\lambda_0 - 0^0$

$$x = 2 \frac{aS \, \Delta\lambda}{\pi} \Bigg\}$$
$$y = aS\phi \qquad\qquad (6.6.3)$$

where λ_0 is the longitude of the central meridian, $\Delta\lambda = \lambda - \lambda_0$, and λ, λ_0, and ϕ are in radians. Again, S is the scale factor.

The grid resulting from Eq. (6.6.3) is shown in Figure 6.2.2. In general it is better to consider the projection as conventional rather than equal area. The distortion at middle latitudes is less than in the equal area

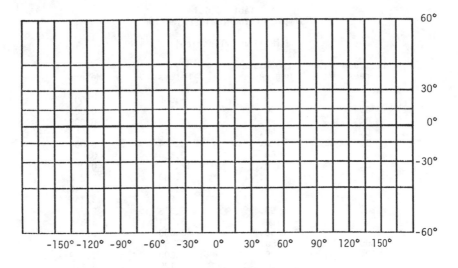

Figure 6.6.1 Perspective Cylindrical Projection

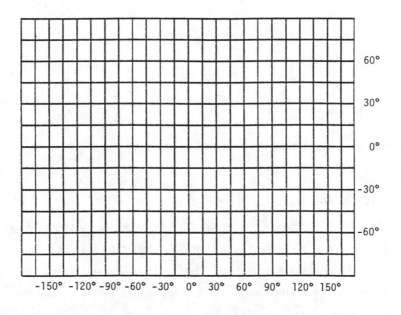

Figure 6.6.2 Miller Cylindrical Projection

Table 6.6.1
Simple Cylindrical Projection

(Miller Cylindrical)

Latitude Degrees	Longitude Degrees	X Meters	Y Meters
0.	0.	.000	0.000
0.	30.	2.126	0.000
0.	60.	4.252	0.000
0.	90.	6.378	0.000
0.	120.	8.504	0.000
0.	150.	10.630	0.000
0.	180.	12.756	0.000
30.	0.	.000	3.340
30.	30.	2.126	3.340
30.	60.	4.252	3.340
30.	90.	6.378	3.340
30.	120.	8.504	3.340
30.	150.	10.630	3.340
30.	180.	12.756	3.340
60.	0.	.000	6.679
60.	30.	2.128	6.679
60.	60.	4.252	6.679
60.	90.	6.378	6.679
60.	120.	8.504	6.679
60.	150.	10.630	6.679
60.	180.	12.756	6.679
90.	0.	.000	10.019
90.	30.	2.126	10.019
90.	60.	4.252	10.019
90.	90.	6.378	10.019
90.	120.	8.504	10.019
90.	150.	10.630	10.019
90.	180.	12.756	10.019

Central meridian: $\lambda_0 = 0^0$

cylindrical, but it is greater at the equator. A plotting table is included
as Table 6.6.1.

6.7 Plate Carreé [24]

The Plate Carreé is a simple cylindrical projection with the equator as
the standard parallel defined by a simple mathematical rule.

The meridians are true length straight lines, parallel to each other.
The meridians are divided as on the sphere, so the parallels are their true
distance apart.   The parallels and the equator are also straight lines,
perpendicular to the meridians. The equator is also divided as on the sphere.

The results of this is the square grid of Figure 6.7.1. The plotting equations which produce this grid are

$$\left. \begin{array}{l} x = aS\Delta\lambda \\ y = aS\phi \end{array} \right\} \cdot \qquad (6.7.1)$$

where $\Delta\lambda = \lambda - \lambda_0$, and λ_0 is the longitude of the central meridian and S is the scale factor. The angles $\Delta\lambda$, and ϕ are in radians.

The distortion in length is extreme along the parallels. The poles, which are points in reality, are represented as straight lines. The projection pretends at neither conformality nor equivalence of area. It does serve as a reasonable diagrammatic representation of data, and is found in many technical reports where no greater cartographic sophistication is required. The Plate Carreé is a map done on standard graph paper.

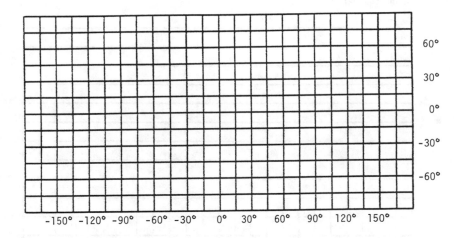

Figure 6.7.1 Plate Carreé Projection

6.8 Carte Parallelogrammatique [24]

The Carte Parallelogrammatique, or die reckteckige plattkarte, has a fancy name for a simple projection. It is essentially a variation for the Plate Carreé, in which two standard parallels, equally spaced around the equator, are taken as true length. The meridians are also true length. The plotting equations, in which ϕ_0 is the latitude of the standard parallels, and λ_0 is the longitude of the central meridian, are

$$\left. \begin{array}{l} x = aS \; \Delta\lambda \; \cos \; \phi_0 \\[2mm] y = aS\phi \end{array} \right\} \; . \tag{6.8.1}$$

in which ϕ, and $\Delta\lambda = \lambda - \lambda_0$ are in radians, and S is the scale factor.

A grid developed from Eq. (6.8.1) is given in Figure 6.8.1. This projection has seen some limited use in atlases. It was developed as a means of reducing some of the distortion inherent in the Plate Carreé. The mapping cylinder is secant to the spherical earth. The area between the standard parallels is smaller, and that poleward from each standard parallel is larger than on the earth.

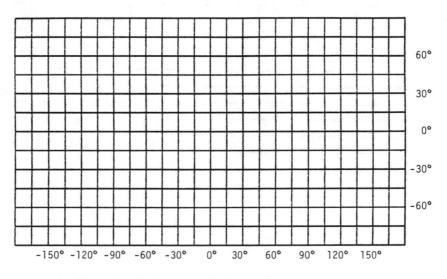

Figure 6.8.1 Carte Parallelogrammetique Projection

6.9 Globular Projection [8], [24]

The globular projection is a convential means of portraying a hemisphere within a circle. In this projection, the central meridian, with longitude λ_0, and one half of the equator are diameters of the circle. The central meridian and the equator are divided truly.

Define a typical circle of parallel of latitude, ϕ. This is shown in Figure 6.9.1. Let d be the distance from the equator to the specified latitude along the central meridian. Let c be the chord length, and h be the distance from the circular arc to the chord. Let ρ be the polar radius vector.

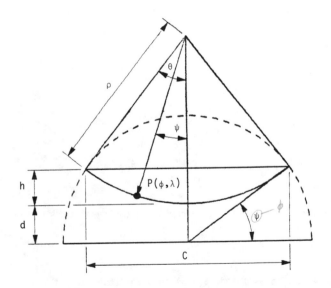

Figure 6.9.1 Geometry for the Globular Projection

The distance along the central meridian, between the equator, and the circle of latitude is

$$d = \frac{a\phi}{\pi/2} \tag{6.9.1}$$

From the figure

$$c = 2a \cos \phi \tag{6.9.2}$$

$$h = a \sin \phi - \frac{a\phi}{\pi/2} \tag{6.9.3}$$

From the geometry of the circular segment

$$c = \sqrt{4h(2\rho - h)} \tag{6.9.4}$$

Substitute Eqs. (6.9.2) and (6.9.3) into (6.9.4), and re-arrange, to obtain the first polar coordinate ρ.

$$c^2 = 4h(2\rho - h)$$

$$2\rho - h = \frac{c^2}{4h}$$

$$2\rho = h - \frac{c^2}{4h}$$

$$\rho = \frac{1}{2} \left(h + \frac{c^2}{4h} \right)$$

$$= \frac{1}{2} \left[a(\sin \phi) \right] - \frac{\phi}{\pi/2} \right] - \frac{4a^2\cos^2\phi}{4a\left[(\sin \phi) - \frac{\phi}{\pi/2}\right]}$$

$$= \frac{a}{2} \left[(\sin \phi) - \frac{\phi}{\pi/2} + \frac{\cos^2 \phi}{(\sin \phi) - \frac{\phi}{\pi/2}} \right] \tag{6.9.5}$$

The angle, θ, also follows from the geometry of the circular segment.

$$\theta = \cos^{-1} \left(\frac{\rho - h}{\rho} \right) \tag{6.9.6}$$

Substitute Eqs. (6.9.3) and (6.9.5) into Eq. (6.9.6)

$$\theta = \cos^{-1} \left\{ \frac{\frac{a}{2}\left[(\sin \phi) - \frac{\phi}{\pi/2} + \frac{\cos 2\phi}{(\sin\phi) - \frac{\phi}{\pi/2}}\right] - a\left[(\sin\phi) - \frac{\phi}{\pi/2}\right]}{\frac{a}{2}\left[(\sin\phi) - \frac{\phi}{\pi/2}\right] + \frac{\cos 2\phi}{(\sin\phi) - \frac{\phi}{\pi/2}}} \right\}$$

$$= \cos^{-1} \left\{ \frac{\cos^2\phi - [(\sin\phi) - \frac{\phi}{\pi/2}]^2}{\cos^2\phi + [(\sin\phi) - \frac{\phi}{\pi/2}]^2} \right\} \tag{6.9.7}$$

The parallels are divided equally. Thus, a second polar coordinate ψ can be defined by the projection, using the relation $\Delta\lambda = \lambda - \lambda_0$

$$\frac{\psi}{\Delta\lambda} = \frac{\theta}{\pi/2}$$

$$\psi = \frac{\Delta\lambda\theta}{\pi/2} \tag{6.9.8}$$

Substitute Eq. (6.9.8) into Eq. (6.9.7).

$$\psi = \frac{2\Delta\lambda}{\pi} \cos^{-1} \left\{ \frac{\cos^2\phi - [(\sin\phi) - \frac{\phi}{\pi/2}]^2}{\cos^2\phi + [(\sin\phi) - \frac{\phi}{\pi/2}]^2} \right\} \tag{6.9.9}$$

The cartesian plotting equations are

$$\left. \begin{array}{l} x = \rho \sin \psi \\ y = d + \rho(1 - \cos \psi) \end{array} \right\} \tag{6.9.10}$$

Substitute Eqs. (6.9.1), (6.9.5) and (6.9.9) into Eq. (6.9.10).

$$x = \frac{a}{2} S[(\sin \phi) - \frac{\phi}{\pi/2} + \frac{\cos^2\phi}{(\sin \phi) - \frac{\phi}{\pi/2}}]$$

$$\sin \left\{ \frac{2\Delta\lambda}{\pi} \cos^{-1} \left\{ \frac{\cos^2\phi - [(\sin\phi) - \frac{\phi}{\pi/2}]^2}{\cos^2\phi + [(\sin\phi) - \frac{\phi}{\pi/2}]^2} \right\} \right\} \tag{6.9.11}$$

$$y = \left\{ aS \frac{\phi}{\pi/2} + \frac{1}{2} [(\sin \phi) - \frac{\phi}{\pi/2} + \frac{\cos^2 \phi}{(\sin \phi) - \frac{\phi}{\pi/2}}] \right\}$$

$$\left\{ 1 - \cos[\frac{2\Delta\lambda}{\pi} \cos^{-1} \left\{ \frac{\cos^2\phi - [(\sin\phi) - \frac{\phi}{\pi/2}]^2}{\cos^2\phi + [(\sin\phi) - \frac{\phi}{\pi/2}]^2} \right\} \right\} \tag{6.9.12}$$

where S is the scale factor. When $\phi = 0^0$, $x = \frac{aS\Delta\lambda}{\pi/2}$, and when $\Delta\lambda = 0^0$, $y = \frac{aS\phi}{\pi/2}$.

Equations (6.9.11) and (6.9.12) produced the grid in Figure 6.9.2. The globular projection has been used for atlas maps.

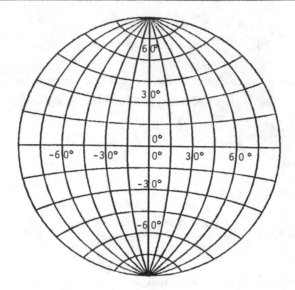

$$60°$$
$$30°$$
$$0°$$
$$-60° \quad -30° \quad 0° \quad 30° \quad 60°$$
$$-30°$$
$$-60°$$

Figure 6.9.2 Globular Projection

6.10 Gall's Projection [24]

Gall's projection is a stereogrphic cylindrical projection, with two standard parallels at 45° north and south. Figure 6.10.1 shows the geometry for the development. The meridians are spaced truly on the two standard parallels. Thus, the abscissa is, with $\Delta\lambda = \lambda - \lambda_0$

$$x = (aS \cos \tfrac{\pi}{4})(\Delta\lambda)$$

$$= 0.70711 \ aS(\Delta\lambda) \qquad\qquad (6.10.1)$$

where λ_0 is the longitude of the central meridian. The ordinates are obtained in a similar manner to the sterographic projection of Section 5.4.

$$y = 1.70711 \ aS \tan \tfrac{\phi}{2} \qquad\qquad (6.10.2)$$

In Eqs. (6.10.11) and (6.10.2), S is the scale factor, and $\Delta\lambda$ is in radians.

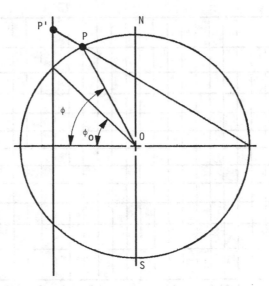

Figure 6.10.1 Gall's Projection

This projection has been successfully used to produce world maps, since the distortion is tolerable. However, it must be kept in mind that neither conformality nor equal area is preserved. The grid for Gall's projection is given as Figure 6.10.2

6.11 Van der Grinten Projection [24]

The Van der Grinten projection contains the complete sphere within a circle. This projection has seen some use in atlas and National Geographic Society maps. While it does not pretend to display conformality or equal area, it does present a pleasing representation of the earth's surface. There is neither the east west extension in higher latitudes that is characteristic of the Mercator, nor the extreme compression in these areas, as shown in the sinusoidal or Mollweide.

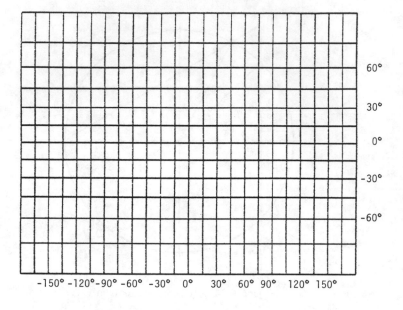

60°

30°

0°

-30°

-60°

-150° -120° -90° -60° -30°   0°   30° 60° 90° 120° 150°

Figure 6.10.2 Gall's Projection

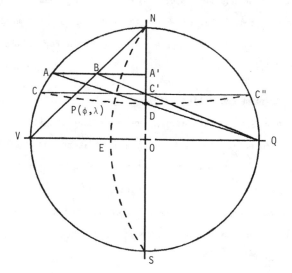

Figure 6.11.1 Geometry for Van der Grinten's Projection

Figure 6.11.1 gives the geometry of the projection. The equator is divided equally, and is represented by the line VQ. The line NS is the central meridian, which is also divided equally.

Consider a purely graphical construction of this projection. Join points N and V. Locate an arbitrary latitude A' on NO, where

$$OA' = NO \frac{\phi}{\pi/2}$$

$$= 2NO \frac{\phi}{\pi} \qquad\qquad (6.11.1)$$

Draw AA' parallel to VQ. The intersection of AA' with NV is B. Join points B and Q. The intersection of BQ and NO defines the point C'. Draw CC' parallel to VQ. Point C constitutes one of the necessary points of the projection. Its symmetric image about NO is a second such point, C". The next step is to connect points A and Q. Then AQ intersects NO at D. This is the third point necessary to completely define a circle of parallel. A circular arc, whose radius is uniquely defined by the location of points C, D, and C" is drawn to obtain a circle of parallel.

Meridians are also circular arcs. These are fit through the poles and the equatorial point for the particular lontigude. From the central meridian, of longitude λ_0, the point on the equator is, with $\Delta\lambda = \lambda - \lambda_0$,

$$EO = VO \frac{\Delta\lambda}{\pi} \qquad\qquad (6.11.2)$$

Modern computer graphics systems usually have the capability of automatically constructing circular arcs given three points [27]. Such a system constructs one arc through points CDC", and a second arc through points NES. The intersection of these arcs gives the location of the desired point P.

6.12 Murdoch's Projection [24]

Murdoch's projection has been used in atlases.   It is a secant projection, but differs from the simple conic with two standard parallels. There are three variations of this projection, but only one of these variations are derived from Figure 6.12.1.

In the first variation, the parallels are spaced their true distance apart on the central meridian, with longitude λ_0. The constant of the cone is

$$c = \sin \left(\frac{\phi_1 + \phi_2}{2}\right) \qquad\qquad (6.12.1)$$

where ϕ_1 is the lower standard parallel, and ϕ_2 is the upper standard parallel.

From the conditions of equal spacing

$$CB = a \left[\frac{2 \sin\left(\frac{\phi_2 - \phi_1}{2}\right)}{\phi_2 - \phi_1}\right] \qquad\qquad (6.12.2)$$

The middle latitude is

$$\psi = \frac{\phi_1 + \phi_2}{2} \qquad\qquad (6.12.3)$$

The radius of the middle parallel is

$$TB = CB \cot \psi \qquad\qquad (6.12.4)$$

The radius of the lower standard parallel is

$$\rho_1 = TB + a(\psi - \phi_1) \qquad\qquad (6.12.5)$$

With the application of Eqs. (6.12.2), (6.12.3) and (6.12.4), and after some simplification, Eq. (6.12.5) becomes

$$\rho_1 = a \left\{\left[\frac{2\sin\left(\frac{\phi_2 - \phi_1}{2}\right)}{\phi_2 - \phi_1}\right]\cos \left(\frac{\phi_1 + \phi_2}{2}\right) + \frac{\phi_2 - \phi_1}{2}\right\} \qquad\qquad (6.12.6)$$

Thus, the first polar coordinate is

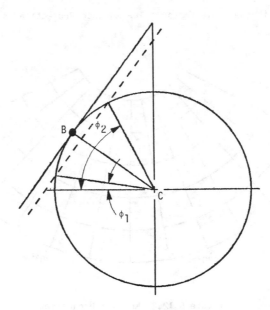

Figure 6.12.1 Geometry for the Murdoch Projection

$$\rho = \rho_1 - a(\phi - \phi_1)$$

$$= a\{[2\sin(\frac{\phi_2 - \phi_1}{2})]\cos(\frac{\phi_1 + \phi_2}{2})$$

$$+ \frac{\phi_2 - \phi_1}{2} - (\phi - \phi_1)\} \qquad (6.12.7)$$

The second polar coordinate is, with $\Delta\lambda = \lambda - \lambda_0$

$$\theta = \Delta\lambda c \qquad (6.12.8)$$

Substitute Eq. (6.12.1) into Eq. (6.12.8).

$$\theta = \Delta\lambda \sin(\frac{\phi_1 + \phi_2}{2}) \qquad (6.12.9)$$

The cartesian plotting coordinates are obtained as

$$\left. \begin{aligned} x &= [\rho \sin \theta]S \\ y &= [\rho_1 - \rho(1 - \cos \theta)]S \end{aligned} \right\} . \qquad (6.12.10)$$

Equations (6.12.5), (6.12.7), (6.12.9) and (6.12.10) produce the desired grid, where S is the scale factor. The Murdoch Projection is shown in Figure 6.12.2.

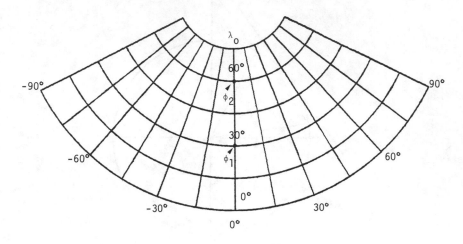

Figure 6.12.2 Murdoch Projection

6.13 Stereographic Variations [24]

Several variations of the stereographic projection have been developed to reduce distortion in regions of particular interest. These are the James, the La Hire, and the Clarke. All of these are geometric perspective projections, and can easily be obtained by an alteration of the stereographic projection of the sphere. Figure 6.13.1 shows the location of the projection points for a polar projections of these variations as compared to those of the gnomonic and the stereogrpahic. La Hire took the projection point as 1.71 times the radius of the earth. James used 1.367 times the radius, and Clarke's two values were 1.35 and 1.65 times the radius. We will consider a derivation of the plotting equations for a polar projection

From Figure 6.13.2,

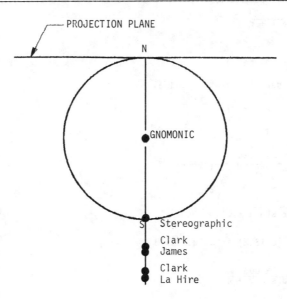

Figure 6.13.1 Projection Points for the Stereographic Variations of the Polar Projections

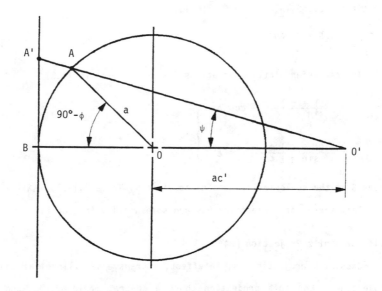

Figure 6.13.2 Geometry for the Stereographic Variations

$$\tan \psi = \frac{\rho}{a(1 + c')}$$

$$\rho = a(1 + c') \tan \psi \qquad\qquad (6.13.1)$$

It is now necessary to relate ψ and ϕ.

$$AB = a \sin(90^0 - \phi)$$

$$= a \cos \phi \qquad\qquad (6.13.2)$$

$$\tan \psi = \frac{AB}{a[\cos(90^0 - \phi) + c']}$$

$$= \frac{AB}{a(\sin \phi + c')} \qquad\qquad (6.13.3)$$

Substitute Eq. (6.13.2) into Eq. (6.13.3)

$$\tan \psi = \frac{a \cos \phi}{a(\sin \phi + c')}$$

$$= \frac{\cos \phi}{\sin \phi + c'} \qquad\qquad (6.13.4)$$

Substitute Eq. (6.13.4) into Eq. (6.13.1).

$$\rho = \frac{a(1 + c')\cos \phi}{\sin \phi + c'} \qquad\qquad (6.13.5)$$

The cartesian plotting equations follow from Eq. (6.13.5)

$$\left. \begin{array}{l} x = \dfrac{aS(1 + c')\cos \phi}{\sin \phi + c'} \cos \Delta\lambda \\[4mm] y = \dfrac{aS(1 + c')\cos \phi}{\sin \phi + c'} \sin \Delta\lambda \end{array} \right\} . \qquad\qquad (6.13.6)$$

where S is the scale factor, and c' is given by Table 6.13.1. $\Delta\lambda = \lambda - \lambda_0$

This series of projections has had some utility in geodetic mapping.

6.14 Cassini's Projection [24]

Cassini's projection is, in effect, a transverse cylindrical equidistant projection. For this projection chose a central meridian, λ_0, and let the origin be at the intersection of this central meridian and the equator. The

Table 6.13.1
Values of c' for the variations of the Stereographic Projection

Name	c'
Clarke	1.35
James	1.367
Clarke	1.65
La Hire	1.71

mapping coordinates follow simply from a consideration of distance on a sphere.

In Figure 6.14.1, P is the arbitrary point on the sphere, with latitude ϕ and longitude λ, and $\Delta\lambda = \lambda - \lambda_0$. PQ is a great circle through P and perpendicular to the central meridian. Let ψ and θ be auxiliary central angles, as shown in the figure. Apply Napier's rules to the spherical triangle to obtain these angles. Recall that $z = 90° - \phi$, which is the colatitude.

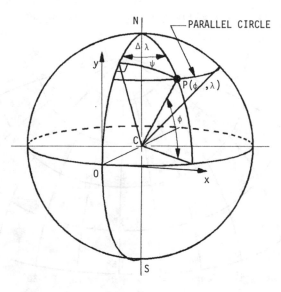

Figure 6.14.1 Geometry for the Cassini Projection

$$\sin \psi = \cos(90^0 - \Delta\lambda)\cos \phi$$

$$= \sin \Delta\lambda \cos \phi$$

$$\psi = \sin^{-1}(\sin \Delta\lambda \cos \phi) \qquad (6.14.1)$$

$$\sin(90^0 - \Delta\lambda) = \tan \phi \tan \theta$$

$$\tan \theta = \frac{\cos \Delta\lambda}{\tan \phi}$$

$$\theta = \tan^{-1} \left(\frac{\cos \Delta\lambda}{\tan \phi}\right) \qquad (6.14.2)$$

The mapping equations follow from Eqs. (6.14.1) and (6.14.2) plus the radius of the sphere.

$$\left. \begin{array}{l} x = a \cdot S \cdot \sin^{-1} (\sin\Delta\lambda \cos \phi) \\[2mm] y = a \cdot S \cdot [\frac{\pi}{2} - \tan^{-1}(\frac{\cos\Delta\lambda}{\tan \phi})] \end{array} \right\} \cdot \qquad (6.14.3)$$

where S is the scale factor.

From Eq. (6.14.3), the true distance on the sphere is maintained, as is also the perpendicularity. Figure 6.14.2 contains the Cassini projection.

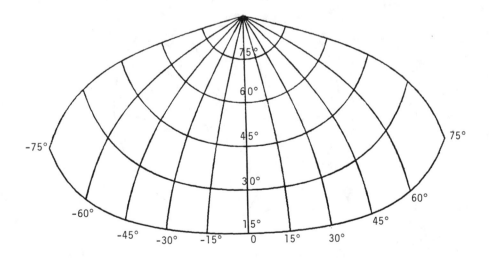

Figure 6.14.2 Cassini's Projection

PROBLEMS

6.1 Given an oblique gnomonic projection with $\phi_0 = 45^0$ and $\lambda_0 = 0^0$.   S:1" = 50,000 meters. Find X and Y if $\lambda = 10°$, and $\phi = 50°$.

6.2 On polar gnomonic projection with $\lambda_0 = 0°$, and S:1" $\equiv$ 50,000 meters, what are the cartesian plotting coordinates if $\lambda = 15°$ and $\phi = 80°$?.

6.3 Given an equatorial gnomonic projection with $\lambda_0 = 0°$, $\lambda = 15°$, and $\phi = 30°$.   S:1" $\equiv$ 100,000 meters.   What are the cartesian plotting coordinates?

6.4 The cartesian plotting coordinates in a polar gnomonic projection are X = 2" and Y = 2:.   S:1" $\equiv$ 100,000 meters.   What are the geographic coordinates?

6.5 The cartesian plotting coordinates on X = 3" and Y = 2", S:1" $\equiv$ 150,000 meters.   Find the geographic coordinates for an equatorial gnomonic projection with $\lambda_0 = 0^0$.

6.6 Repeat Problem 6.1 for an azimuthal equidistant oblique projection.

6.7 Repeat Problem 6.2 for an azimuthal equidistant polar projection.

6.8 Repeat Problem 6.3 for an azimuthal equidistant equatorial projection.

6.9 Repeat Problem 6.1 for an oblique orthographic projection.

6.10 Repeat Problem 6.2 for a polar orthographic projection.

6.11 Repeat Problem 6.3 for an equatorial orthographic projection.

6.12 Given a simple conic with one standard parallel at latitude 45°, $\lambda_0 = 0°$, S:1" $\equiv$ 50,000 meters.   For $\phi = 35°$, and $\lambda = 50°$, find the cartesian coordinates.

6.13 For a simple conic with two standard parallels at latitudes 30° and 60°, λ_0, = 0°, S:1" $\equiv$ 50,000 meters. What are the cartesian coordinates when $\phi = 40°$, and $\lambda = 25°$.

6.14 Repeat Problem 6.13 when $\phi = 70°$, and $\lambda = 25°$.

6.15 Repeat 6.12 for perspective conic.

6.16 For a latitude of 35° and a longitude of 25°, what are the cartesian plotting coordinates on a polyconic projection? Let $\phi_0 = 15°$, and $\lambda_0 = 0°$.   S:1" $\equiv$ 150,000 meters.

6.17 Drive the transverse polyconic direct transformation equations.

6.18 Derive the inverse transformation equations for the regular polyconic projection.

6.19 Given a perspective cylindrical projection with $\lambda_0 = 0°$, and S:1" $\equiv$ 500,000 meters.   What are the cartesian plotting coordinates if $\phi = 30°$, and $\lambda = -30°$?

6.20 Repeat Problem 6.19 for the Miller cylindrical projection.

6.21 Repeat Problem 6.19 for the Plate Carreé.

6.22 Repeat Problem 6.19 for the Carte Parallelogrammatique with ϕ_0 = 20°.

6.23 Given the globular projection with S:1" ≡ 100,000 meters. If ϕ = 35°, and λ = 35°, what are the cartesian plotting coordinates?

6.24 Repeat Problem 6.19 for Gall's projection.

6.25 Repeat Problem 6.13 for Murdoch's projection.

6.26 Let λ_0 = 0°.   For ϕ = 30°, and λ = 20°, calculate the cartesian coordinates for the two Clarke stereographic variations.

6.27 Repeat Problem 6.26 for the James and La Hire stereographic variations.

6.28 Repeat Problem 6.23 for Cassini's projection.

7

Theory of Distortions

Distortion is the cartographic bane. This was mentioned in Chapter 1. No matter what technique or algorithm is used, distortion does occur in length, angle or area, or in a combination of these. Throughout Chapters 4, 5, and 6, it was pointed out where distortions do occur for particular projections, but this was treated only qualitatively. This brief chapter deals with the general theory of distortion in maps in a quantitative way. Formulas are developed to quantify the distortions in length, angle, and area as introduced in Chapter 1. Then, these generalities are applied to the most frequently used equal area, conformal, and conventional projections. Thus, we suggest a numerical means of assessing the acceptability of a map for a particular application. Finally, some of the different classes of maps are compared in a qualitative manner.

7.1 Distortion in Length [10], [22]

Dealing with distortions in length requires consideration of changes in local scale as defined in Chapter 1. In order to describe these distortion in length, we consider a two dimensional plotting surface, and derive terms for distortion along the parallels and meridians, as compared to true distance along a sphere or spheroid. The derivation begins with the first fundamental forms of the earth and the plotting surface. The ratio of these fundamental forms is defined to be m. From Eq. (2.3.3)

$$m^2 = \frac{E(d\phi)^2 + 2 F \, d\phi \, d\lambda + G(d\lambda)^2}{e(d\phi)^2 + 2 f \, d\phi \, d\lambda + g(d\lambda)^2} \tag{7.1.1}$$

where the capital letters refer to the mapping surface, and the lower case, to the sphere or spheroid. Since we are dealing exclusively in orthogonal systems for the mapping surface and the earth, the substitution of $F = f = 0$

into Eq. (7.1.1) yields

$$m^2 = \frac{E(d\phi)^2 + G(d\lambda)^2}{e(d\phi)^2 + g(d\lambda)^2} \tag{7.1.2}$$

The distortion along the parametric ϕ-curve, or meridian, where $d\lambda = 0$, is from Eq. (7.1.2).

$$m_m = \sqrt{\frac{E}{e}} \tag{7.1.3}$$

and the distortion along the parametric λ-curve, perpendicular to the meridian, where $d\phi = 0$, is

$$m_p = \sqrt{\frac{G}{g}} \tag{7.1.4}$$

These distortions in distance will be applied to particular projections in Sections 7.4, 7.5, and 7.6. True length corresponds to $m_p = m_m = 1$. These quantities, m_p and m_m are the indicators of local scale.

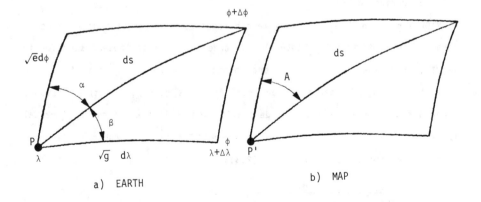

a) EARTH   b) MAP

Figure 7.2.1 Differential Parallelograms

7.2 Distortions in Angles [10], [22]

Figure 7.2.1a is needed for the derivation of the angular distortion. Again, from the first fundamental for the model of the earth

$$(ds)^2 = e(d\phi)^2 + 2f(d\phi d\lambda) + g(d\lambda)^2. \qquad (7.2.1)$$

The angle between the parametric ϕ-, and λ-curves intersecting at point P is

$$\omega = \alpha + \beta \qquad (7.2.2)$$

Consider the differential parallelogram to be sufficiently small in area, so that it can be treated as a plane. Then, the law of cosines applies.

$$(ds)^2 = e(d\phi)^2 + g(d\lambda)^2 + 2\sqrt{eg}\, d\phi\, d\lambda\, \cos\omega \qquad (7.2.3)$$

Equatiing Eqs. (7.2.1) and (7.2.3)

$$\cos\omega = \frac{f}{\sqrt{eg}} \qquad (7.2.4)$$

Also, we will have

$$\sin\omega = \sqrt{\frac{eg - f^2}{eg}} \qquad (7.2.5)$$

Since we will deal with orthogonal systems, f = 0, and from Eqs. (7.2.4) and (7.2.5)

$$\cos\omega = 0$$
$$\sin\omega = 1$$
$$\omega = \pi/2$$

Thus, from Eq. (7.2.2)

$$\alpha + \beta = \pi/2 \qquad (7.2.6)$$

From the figure

$$\left.\begin{array}{l} \cos\alpha = \sqrt{e}\,\dfrac{d\phi}{ds} \\[2mm] \sin\alpha = \sqrt{g}\,\dfrac{d\lambda}{ds} \end{array}\right\} \qquad (7.2.7)$$

Consider now, Figure 7.2.1b, where an infinitesimal part of the projection surface is shown at point P', corresponding to point P on the earth. From the figure, for the orthogonal mapping surface,

$$\left.\begin{array}{l} \cos A = \sqrt{E}\,\dfrac{d\phi}{dS} \\[2ex] \sin A = \sqrt{G}\,\dfrac{d\lambda}{dS} \end{array}\right\} \qquad\qquad (7.2.8)$$

Expand

$$\sin(A - \alpha) = \sin A \cos \alpha - \cos A \sin \alpha \qquad\qquad (7.2.9)$$

Substitute Eqs. (7.2.7) and (7.2.8) into Eq. (7.2.9).

$$\sin(A - \alpha) = \sqrt{Ge}\,\frac{d\lambda}{dS}\frac{d\phi}{ds} - \sqrt{Eg}\,\frac{d\lambda}{dS}\frac{d\phi}{dS}$$

$$= (\sqrt{Ge} - \sqrt{Eg})\,\frac{d\phi}{ds}\frac{d\lambda}{dS} \qquad\qquad (7.2.10)$$

Substitute Eqs. (7.1.3) and (7.1.4) into Eq. (7.2.10).

$$\sin(A - \alpha) = (m_p - m_m)\,\frac{d\phi}{dS}\frac{d\lambda}{dS}\,\sqrt{eg} \qquad\qquad (7.2.11)$$

In a similar expansion,

$$\sin(A + \alpha) = (m_p + m_m)\,\frac{d\phi}{dS}\frac{d\lambda}{dS}\,\sqrt{eg} \qquad\qquad (7.2.12)$$

Re-arrange Eq. (7.2.11) and (7.2.12), and equate

$$\sin(A - \alpha) = \frac{m_p - m_m}{m_p + m_m}\,\sin(A + \alpha) \qquad\qquad (7.2.13)$$

Equation (7.2.13) is another transcendental beast. For a constant m_p and m_m it can be solved by the Newton-Raphson method [14]. Write Eq. (7.2.13) as

$$f(\alpha) = 0$$

$$= \sin(A - \alpha) - \frac{m_1 - m_m}{m_p + m_m}\,\sin(A + \alpha) \qquad\qquad (7.2.14)$$

The derivative of Eq. (7.2.14) is

$$\frac{df}{dA} = \cos(A - \alpha) - \frac{m_p - m_m}{m_p + m_m}\,\cos(A - \alpha) \qquad\qquad (7.2.15)$$

The Newton-Raphson scheme for the solution of Eq. (7.2.13) is then

$$A_{n+1} = A_n - \frac{f}{\frac{df}{dA}}. \tag{7.2.16}$$

Substitute Eqs. (7.2.14) and (7.2.15) into Eq. (7.2.16).

$$A_{n+1} = A_n$$

$$- \frac{\sin(A_n - \alpha) - (\frac{m_p - m_m}{m_p + m_m})\sin(A_n + \alpha)}{\cos(A_n - \alpha) - (\frac{m_p - m_m}{m_p + m_m})\cos(A_n + \alpha)} \tag{7.2.17}$$

As an intialization, let $A_0 = \alpha$. This iteration is rapidly convergent, and easily computerized.

This technique is applied to the equal area, and the conventional projections. As is seen in Section 7.6, Eq. (7.2.13) has a unique solution for conformal projections.

7.3 Distortion in Area [10], [22]

The area on the map is, from (2.3.13)

$$A_m = EG - F^2 \tag{7.3.1}$$

and the area on the model of the earth is

$$A_e = eg - f^2 \tag{7.3.2}$$

The distortion of area is hereby defined to be

$$D_A = A_m/A_e$$

$$= \sqrt{\frac{EG - F^2}{eg - f^2}} \tag{7.3.3}$$

Since the systems are orthogonal, $F = f = 0$ can be substituted into Eq. (7.3.3).

$$D_A = \sqrt{\frac{EG}{eg}} \tag{7.3.4}$$

Substitute Eqs. (7.1.3) and (7.1.4) into Eq. (7.3.4).

$$D_A = \sqrt{m_p m_m} \tag{7.3.5}$$

Now we are in a position to consider these distortions in terms of selected projections. In what follows, the distortions are derived for polar and regular cases. They apply equally for the oblique, transverse, and equatorial cases. The only difference is that α is substituted for λ, and h is substituted for ϕ.

7.4 Distortion in Equal Area Projections [32]

Equal area projections, by their definition, have no distortion in area in the mapping transformation. Thus, from Eq. (7.3.5)

$$D_A = \sqrt{m_p m_m}$$

$$= 1 \tag{7.4.1}$$

It is seen from Eq. (7.4.1) that m_p and m_m are the reciprocals of each other. We now consider the more important equal area projections of Chapter 4, the conical with one standard parallel, and with two standard parallels, the polar azimuthal, and the cylindrical. The distortions in length and angle, at an arbitrary point, are derived. In the case of distortions in length, only one of the distortion factors needs to be derived. Equation (7.4.1) is then used to find the second one. Thus, in all cases, the easier of the two will be chosen for the derivation, and the second is its reciprocal. All cases are derived by considering the authalic sphere of Section 4.1.

Consider first the conical projection with one standard parallel, the first of the Albers projections. From Eqs. (7.1.3), (4.2.1), (4.2.2), and (4.2.5)

$$m_m = \sqrt{\frac{E}{e}}$$

$$= \sqrt{\frac{c_1^2 \rho^2}{R^2 \cos^2 \phi}}$$

$$= \frac{c_1 \rho}{R \cos \phi} \tag{7.4.2}$$

Substitute Eqs. (4.2.15) and (4.2.19) into Eq. (7.4.2.

$$m_m = \frac{\sin \phi_0 \, \frac{R}{\sin \phi_0} \, \sqrt{1 + \sin^2 \phi_0 - 2 \sin \phi \sin \phi_0}}{R \cos \phi}$$

$$= \frac{\sqrt{1 + \sin^2 \phi_0 - 2 \sin \phi \sin \phi_0}}{\cos \phi} \tag{7.4.3}$$

From eq. (7.4.1)

$$m_p = \frac{\cos \phi}{\sqrt{1 + \sin^2 \phi_0 - 2 \sin \phi \cos \phi_0}} \tag{7.4.4}$$

It is obvious from Eqs. (7.4.3) and (7.4.4) that an expansion in scale in one direction is offset by a contraction in the direction orthogonal to it.

Consider next the case of the two standard parallel Albers projection. From Eqs. (4.2.6), (4.2.32) and (7.4.2)

$$m_m = \frac{(\sin \phi_1 + \sin \phi_2)}{2R \cos \phi} \sqrt{\rho_2^2 + \frac{4R^2 (\sin \phi_2 - \sin \phi)}{\sin \phi_1 + \sin \phi_2}} \tag{7.4.5}$$

From Eq. (4.2.33) we have

$$m_m = \frac{(\sin \phi_1 + \sin \phi_2)}{2R \cos \phi} \sqrt{\frac{4R^2 \cos^2 \phi + 4R^2 (\sin \phi_2 - \sin \phi)}{\sin \phi_1 + \sin \phi_2}}$$

$$= \sqrt{1 + \frac{(\sin \phi_1 + \sin \phi_2)(\sin \phi_2 - \sin \phi)}{\cos^2 \phi}} \tag{7.4.6}$$

Then, from (7.4.1) is obtained

$$m_p = \sqrt{\frac{\cos^2 \phi}{\cos^2 \phi + (\sin \phi_1 + \sin \phi_2)(\sin \phi_2 - \sin \phi)}} \tag{7.4.7}$$

Next, we consider the polar azimuthal, or Lambert, projection. In this case, $\sin \phi_0 = 1$, is substituted into Eq. (7.4.2)

$$m_m = \frac{\rho}{R \cos \phi} \tag{7.4.8}$$

Substitute Eq. (4.3.2) into Eq. (7.4.8)

$$m_m = \frac{R \sqrt{2(1 - \sin \phi)}}{R \cos \phi}$$

$$= \frac{\sqrt{2(1 - \sin \phi)}}{\cos \phi} \qquad (7.4.9)$$

From (7.4.1)

$$m_p = \frac{\cos \phi}{\sqrt{2(1 - \sin \phi)}} \qquad (7.4.10)$$

The cylindrical equal area projection was treated in Section 4.5. From Eq. (7.1.3), (4.5.2), and an application of the fundamental transformation matrix

$$m_m = \sqrt{\frac{E}{e}}$$

$$= \sqrt{\frac{R^2}{R^2 \cos^2 \phi}}$$

$$= \frac{1}{\cos \phi} \qquad (7.4.11)$$

From (7.4.1)

$$m_p = \cos \phi \qquad (7.4.12)$$

Note that the four sets of distortion factors, Eqs. (7.4.3) and (7.4.4), Eqs. (7.4.6) and (7.4.7), Eqs. (7.4.9) and (7.4.10), and (7.4.11) and (7.4.12) ae all independent of longitude. Thus, in these cases, distortion in length if a function of latitude alone.

In order to find distortion in angles, substitute Eq. (7.4.1) into Eq. (7.2.13).

$$\sin(A - \alpha) = \frac{(m_p - \frac{1}{m_p})}{(m_p + \frac{1}{m_p})} \sin(A - \alpha)$$

$$\sin(A - \alpha) = (\frac{m_p^2 - 1}{m_p^2 + 1}) \sin(A + \alpha) \qquad (7.4.13)$$

Equation (7.4.13) can now be solved for constant m_p by the iteration method of Section 7.2.

As an example, consider a polar azimuthal projection at $\phi = 60°$

$$m_p = \frac{\cos 60°}{\sqrt{2}(1 - \sin 60°)} = \frac{0.500}{\sqrt{(2)}(1 - 0.866)} = 0.965$$

$$m_m = \frac{1}{m_p} = \frac{1}{0.965} = 1.036$$

7.5 Distortion in Conformal Projections [22]

From Chapter 5, it is recalled that conformal projections are characterized by the fact that

$$m^2 = \frac{E}{e} = \frac{G}{g} \qquad (7.5.1)$$

for an orthogonal system. Thus, from Eqs. (7.1.3) and (7.1.4), at every point

$$m_p = m_m \qquad (7.5.2)$$

This relationship makes the work involved in any derivation of linear distortions easier, since only one ratio of the first fundamental quantities needs to be evaluated. In this section, the polar stereographic, the Lambert conformal with one and two standard parallels, and the equatorial Mercator projection are considered. These distortions are based on the spheroid as the earth surface.

For the Lambert conformal projection with one standard parallel, using (7.1.4)

$$m = m_p$$

$$= m_m$$

$$= \sqrt{\frac{G}{g}} \qquad (7.5.3)$$

Substitute Eqs. (5.3.2) and (5.3.3) into Eq. (7.5.3).

$$m = \sqrt{\frac{\rho^2}{R_p^2 \cos^2 \phi}}$$

$$= \frac{\rho}{R_p \cos \phi} \qquad (7.5.4)$$

Substitute Eqs. (5.3.14) and (5.3.21) into Eq. (7.5.4).

$$m = R_{po} \cot \phi_0$$

$$\cdot \left\{ \frac{\tan(\frac{\pi}{4} - \frac{\phi_1}{2})(\frac{1 + e \sin \phi}{1 - e \sin \phi})^{e/2}}{\tan(\frac{\pi}{4} - \frac{\phi_0}{2})(\frac{1 + e \sin \phi_0}{1 - e \sin \phi_0})^{e/2}} \right\}^{\sin \phi_0} \frac{1}{R_p \cos \phi} \qquad (7.5.5)$$

For the two standard parallel case, the distortion in length is obtained
in a manner similar to the one standard parallel case. From Eq. (7.5.4)

$$m = \frac{\rho \sin \phi_0}{R_p \cos \phi} \qquad (7.5.6)$$

Equation (7.5.6) is then evaluated with the aid of Eqs. (5.3.28), (5.3.31),
and (5.3.32).

For the stereogrpahic polar projection, the linear distortion along the
parametric curves may be found by considering the plane and the spheroid.
From the polar coordinates of the mapping plane

$$G = \rho^2 \qquad (7.5.7)$$

For the spheroid, again

$$g = R_p^2 \cos^2 \phi \qquad (7.5.8)$$

Substituting Eqs. (7.5.7) and (7.5.8) into Eq. (7.5.1) we obtain

$$m = \frac{\rho}{R_p \cos \phi} \qquad (7.5.9)$$

Substitute Eq. (5.4.6) into Eq. (7.5.9)

$$m = \frac{2a}{\sqrt{1 - e^2}} (\frac{1 - e}{1 + e})^{e/2} \tan(\frac{\pi}{4} - \frac{\phi}{2})$$

$$\cdot (\frac{1 + e \sin \phi}{1 - e \sin \phi})^{e/2} (\frac{1}{R_p \cos \phi}) \qquad (7.5.10)$$

As a final effort in this section, we will consider the equatorial
Mercator projection. Again, the spheroid has the first fundamental quantity

$$g = R_p^2 \cos^2 \phi \qquad (7.5.11)$$

For the plotting surface

$$G = a^2 \qquad\qquad (7.5.12)$$

Substitute Eqs. (7.5.11) and (7.5.12) into Eq. (7.5.1)

$$m = \frac{a}{R_p \cos \phi} \qquad\qquad (7.5.13)$$

Consider the linear distortions given in Eqs. (7.5.5), (7.5.6), (7.5.10), and (7.5.13). Again, all of these equations depend only on latitude. In any practical evaluation of these distortions, it is sufficient to treat the earth as a sphere, rather than as a spheroid. This is done by letting e = 0 in the equations for the distortions. These equations then reduce to easily manageable forms. As an example, $R_p = a$, and Eq. (7.5.13) for the equatorial Mercator becomes $m = \frac{1}{\cos \phi}$. Evaluated at $\phi = 30°$

$$m = \frac{1}{0.866} = 1.155$$

For the distortion in angle at a point, substitute Eq. (7.5.2) into Eq. (7.2.13)

$$\sin(A - \alpha) = \frac{(m_p - m_p)}{(m_p + m_p)} \sin(A + \alpha)$$

$$= 0$$

$$A - \alpha = 0$$

$$A = \alpha \qquad\qquad (7.5.14)$$

Thus, one of the properties of the conformal projection is that angles are preserved in the transformation.

7.6 Distortions in Conventional Projections [1], [22]

Three of the most important conventional projections will be considered. These are the gnomonic, the azimuthal equidistant, and the polyconic. In these projections, $m_p \neq m_m$, and $m_p \neq 1/m_m$. Since there exists no simple relation between the two distortions in length along the parametric curves, it is necessary to solve for both. However, basic definitions make this simple in many cases.

Consider first the polar gnomonic projection. From the geometry of the case (Figure 6.1.1), where δ is the colatitude or $\delta = 90^0 - \phi$,

$$m_m = \frac{R \tan \delta \, d\lambda}{R \sin \delta \, d\lambda}$$

$$= \frac{1}{\cos \delta}$$

$$= \frac{1}{\sin \phi} \qquad\qquad (7.6.1)$$

For the distortion perpendicular to the meridian

$$m_p = \frac{dS}{ds}$$

$$= \frac{1}{\sin^2 \phi} \qquad\qquad (7.6.2)$$

For the azimuthal equidistant polar projeciton, the distance along the meridians is true, by definition. Thus,

$$m_m = 1 \qquad\qquad (7.6.3)$$

The distance along the parallels on the map, as compared to those on the sphere follows from the geometry of a circular segment.

$$m_p = \frac{2 \pi_a (\frac{\pi}{2} - \phi)}{2 \pi a \cos \phi}$$

$$= \frac{\pi/2 - \phi}{\cos \phi} \qquad\qquad (7.6.4)$$

The last projection to be considered is the regular polyconic. By the assumptions included in the derivation of the projection

$$m_p = 1 \qquad\qquad (7.6.5)$$

The distortion along the meridians is given by

$$m_m = 1 + \frac{\lambda^2}{2} \cos^2 \lambda \qquad [1] \qquad\qquad (7.6.6)$$

For the gnomonic and azimuthal equidistant projections, the linear distortion is independent of longitude. However, the distortion in the meridian plane for the polyconic is a function of both latitude and longitude. As an example, consider a polar gnomonic projection with $\phi = 60°$

$$m_m = \frac{1}{\sin 60°} = \frac{1}{0.866} = 1.155$$

$$m_p = \frac{1}{\sin^2 60°} = \frac{1}{(0.866)^2} = 1.333$$

Distortion in angles at a point must be found by the numerical technique of Eq. (7.2.17).

7.7 Qualitative Comparisons [8], [21]

While a numerical approach is certainly useful, qualitative comparisons of the various projections show dramatically the effects of distortion. This section compares selected azimuthal, world, cylindrical, and conical projections.

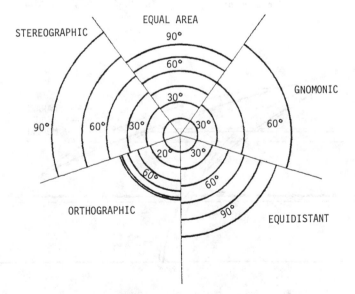

Figure 7.7.1 Comparison of Azimuthal Polar Projections.

Figure 7.7.1 compares five of the azimuthal polar projections: the equal area, the equidistant, the orthographic, the stereographic, and the gnomonic. Beginning with the orthographic, there is a steady gradation of parallel spacing, ending with the gnomonic. The orthographic projection suffers distortion as the equator is approached. The parallels are unequally spaced, and are bunched together close to the equator. The convergence of the parallels is not as severe for the equal area projection. The equidistant projection has equally spaced parallels. The stereographic and gnomonic projections have a divergence of the concentric parallels as the equator is approached. The distortion in the gnomonic projection is more severe, and the equator itself can never be protrayed.

In Figure 7.7.2, one quadrant of three equal area world maps at the same scale and radius is plotted. Recall from the repective plotting equations for these projections that each quadrant has an area of πR^2. The sinusoidal projection results in a pointed figure. The Mollweide projection is a smooth

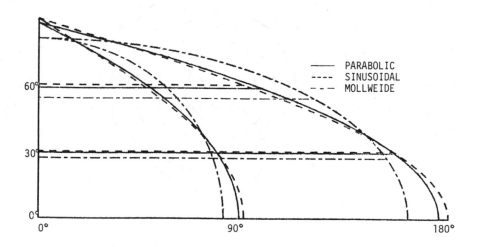

Figure 7.7.2 Comparison of Equal Area World Maps.

curve with a shorter semi-equator. The parabolic projection is intermediate in terms of curvature and length of the semi-equator. Note also the variation in the spacing of the parallels.

In Figure 7.7.3 is a comparison of the Mercator, the Plate Careé, and the cylindrical equal area, all cylindrical projections. The Plate Careé has equal spacing along the meridian. The spacing for the Mercator increases as higher latitudes are reached. This is reversed in the equal area cylindrical. Here, the spacing decreases at higher latitudes.

Three conical projections of one standard parallel are compared in Figure 7.7.4. These are the equal area, the perspective, and the Lambert conformal. In the equal area projeciton, the parallels are closer at higher latitudes, and further at lower latitude. In the Lambert conformal, the spacing diverges north and south of the standard parallel, but in a gradual way. In the perspective projection, the divergence is far more severe.

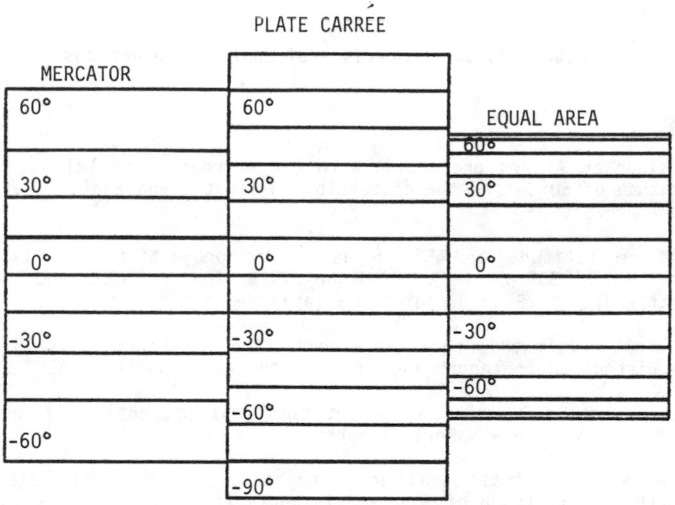

Figure 7.7.3 Comparison of Cylindrical Projections

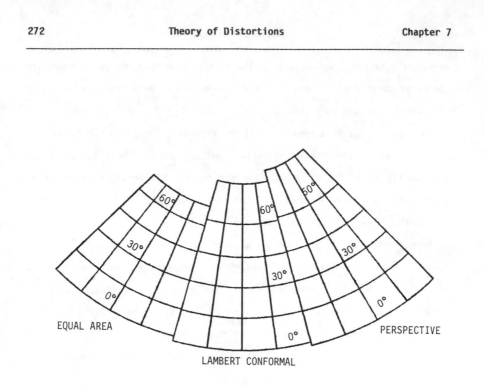

Figure 7.7.4 Comparison of Conical Projections

PROBLEMS

7.1 Consider an Albers projection with one standard parallel at 45°. For a latitude of 50°, find the distortion in length, and angle. Let the earth angle be 75°.

7.2 Let the latitude be 45° on an Albers projection with two standard parallels at 30° and 60°. Find the distortion in length and angle if the earth angle is 75°. Repeat for a latitude of 65°.

7.3 On a polar azimuthal equal area projection the latitude is 75°. What is the distortion in length and angle if the earth angle is 45°?

7.4 Repeat Problem 7.1 for a Lambert conformal projection of one standard parallel. Assume a spherical earth.

7.5 Given a cylindrical equal area projection, what is the distortion in length at a latitude of 35°, and in angle if the earth angle is 60°?

7.6 Repeat Problem 7.2 for a Lambert conformal projection with two standard parallels. Assume a spherical earth.

7.7 Repeat Problem 7.3 for a polar stereographic projection and a spherical earth.

7.8 Repeat Problem 7.5 for a Mercator projection and a spherical earth.

7.9 Repeat Problem 7.3 for a polar gnomonic projection.

7.10 Repeat Problem 7.3 for a polar azimuthal equal area projection.

7.11 Given a polyconic projection. Find the distortion in length and angle at latitude 45°, and change in longitude of 45°. Let the earth angle be 60°.

8

Uses of Map Projections

This concluding chapter considers recommended uses for particular map projections. As a tool in this effort it is useful to give a guide to the identification of map projections based on the characteristics of the meridians and parallels. Then, in the final section of the chapter, a brief summary is given of the most important points covered on the way to understanding the methods of map projection, and applying them as an art.

8.1 Characteristics of the Projections

The obvious basis for the identification of map projections is the characteristics of the meridians and parallels in each particular case. Tables 8.1.1 and 8.1.2 present the means of identification of projections with straight meridians, and curved meridians, respectively.

Table 8.1.1 considers projections with straight meridians either parallel, radial, or converging. The parallels are either straight lines, concentric circles, or concentric arcs. The relative spacing of the parallels is a key to the identification of the projections.

In Table 8.1.2, the first criterion is whether or not the curved meridians are equally spaced along a given parallel. The parallels may be arcs, either concentric or not, or straight lines, equally spaced or not.

8.2 Recommended Uses of Map Projections

Associated with the theory and practice of map projection methods is the need to use map projections correctly. It must be kept in mind that the use of map projections is an art. The rule, of course, is to choose a system that minimizes distortion to an acceptable level in the region of interest.

Consider first two of the major categories of map projections, the equal area and the conformal projections. The equal area projections, as a class, are excellent for the accurate display of statistical data [28]. The equal area property faithfully maintains proper proportions on the map. However, this is purchased at sometimes extreme distortions in shape. On the other

Table 8.1.1
Identification of Map Projections with Straight Line Meridians

Meridians	Parallels	Projection
Parallel	Straight lines	
	Widest apart at equator	Cylindrical Equal Area
	Equidistant	Simple cylindrical
	Widest apart at poles	Perspective cylindrical
		Mercator
Radial	Concentric circles	
	Widest apart away from poles	Polar gnomonic
		Polar stereographic
	Equidistant	Polar azimuthal equidistant
	Widest apart at poles	Polar azimuthal equal area
		Polar orthographic
Converging	Arcs concentric	Simple conics
		Equal area conics
		Lambert conformal

hand, the conformal projections maintain shape relatively well, but at the expense of size. The use of conformal projections to display statistical data brings about a misleading impression.

The precept for choosing the correct map for the required job is helped by the apparent natural adaptability of types of projections to certain areas of the earth. By consulting the projection tables and the figures, it is seen, that the azimuthal polar projections can easily, and with minimum distortion, handle areas immediately adjacent to the earth's poles. Likewise, the cylindrical projections are natural for the regions above and below the equator. The areas at mid-latitudes are conveniently spanned by conical projections of one or two standard parallels. Recall that in these projections, distortion is not a function of longitude. Thus, for a conformal

Table 8.1.2
Identification of Map Projections with Curved Meridians

Meridians	Parallels	Projection
Equally spaced along a given parallel	Arcs	
	Concentric	Bonne Werner
	Not concentric	Polyconic Hammer-Aitoff
	Straight lines Equally spaced	Sinusoidal
	Not equally spaced	Mollweide Parabolic Eumorphic
Not equally spaced along a given parallel	Straight lines	Equatorial orthographic Oblique & Equatorial Azmuthal Equal Area
	Arcs	Transverse Mercator Oblique & Equatorial Stereographic Oblique & Equataorial Gnomonic Oblique Azimuthal Equidistant Globular

representation of the world, a natural set is the Mercator between ± 30° latitude, a Lambert conformal from ± 30° to 60°, and the polar stereographic from ± 60° to ± 90°. For an equal area approach, the same regions can be covered by an equal area cylindrical, an Albers', and the Lambert azimuthal, respectively.

If there is a desire to include an entire sphere or a hemisphere on a map, the polyconic, or the world statistical maps are the best approach. However, one must learn to live with the excessive distortion, or use interrupted variations. Note that only equal area and conventional projections can be used for a map of the entire earth. Extreme distortions at the periphery of conformal projections will not permit this.

Specific applications have called for the utilization of specific projections. Air and sea navigation have required the Mercator with the loxodrome, and the gnomonic, with the great circle over extended distances.

Intermediate distances, such as a number of states or less, has fallen in the province of the polyconic and the Lambert conformal. In both of these projections, excessive distortion is not evident. Surveying systems have made use of the Transverse Mercator and Lambert conformal on large scale maps. Here, the curvature of the meridians and parallels are not noticeable. If one needs true distance and azimuths from a fixed point to anywhere in the world, the oblique azimuthal equidistant can accomodate. For large scale maps, and relatively short distances, any of the conformal maps, and also the Bonne projection, can be used for relatively good approximations of distance and azimuth.

Another consideration relates to the accuracy required. Some applications require considerable accuracy, such as in navigation, or state plane coordinate calculations. Other applications do not require as much accuracy to convey the necessary information. As an example of the latter, consider an interactive computer graphics system in which a map is digitized, and selected points are transformed from cartesian to geogrpahic coordinates. There is a limit on the accuracy obtained in this transformation. The resolution of the digitizing, and imperfections in the map due to such things as humidity, or methods of reproduction will limit the accuracy of the final result. A spherical model of the earth, with a mean radius is probably all that is required. Only when extreme accuracy is required is it necessary to use the appropriate spheroidal model. The user must keep his required level of accuracy in mind.

Another consideration is the display of enumerative data. In contrast to the display of statistical data, where relative area is important, one can wish to display enumerative data, where only the number of points is important. Then any one of the projections in this text can be arbitrarily chosen.

And finally, keep in mind that maps are often used as a decorative note. More often in this case, the more exotic looking projections, much as Werner, the interrupted projections, or the equal area world maps become elements in interior decorating, or a company logo.

Tables 8.2.1, 8.2.2, and 8.2.3 are included to summarize recommended regions of coverage for equal area, conformal, and conventional projections, respectively. The tables are marked with an "X" for those regions that are best portrayed on a particular projection. The regions that are considered are the polar, the mid latitudes, the equatorial, a transverse zone, and the entire world. Those projections that can be made applicable to a region by a transverse rotation are marked by a "T".

Table 8.2.1
Regions of Coverage of Equal Area Projections

Projection	Polar	Mid Latitude	Equatorial	Transverse Zone	World
Conical, 1 standard parallel		X			
Conical, 2 standard parallels		X			
Polar azimuthal	X				
Oblique azimuthal		X			
Equatorial azimuthal			X		
Bonne		X			
Cylindrical			X	T	
Mollweide			X		X
Parabolic			X		X
Hammer-Aitoff			X		X
Eumorphic			X		X

Table 8.2.2
Regions of Coverage of Conformal Projections

Projection	Polar	Mid Latitude	Equatorial	Transverse Zone	World
Equatorial Mercator			X		
Oblique Mercator		X	X		
Transverse Mercator				X	
Lambert conformal, 1 standard parallel		X			
Lambert conformal, 2 standard parallels		X			
Polar stereographic	X				
Oblique stereographic		X			
Equatorial stereographic			X		

Table 8.2.3
Regions of Coverage of Conventional Projections

Projection	Polar	Mid Latitude	Equatorial	Transverse Zone	World
Polar gnomonic	X				
Oblique gnomonic		X			
Equatorial gnomonic			X		
Oblique azimuthal equadistant		X			
Polar azimuthal equidistant	X				
Polar orthographic	X				
Equatorial orthographic			X		
Simple conic variation		X			
Polyconic		X	X	X	
Simple cylindrical variations			X	T	
Plate Carree			X		X
Carte Parallelogrammatique			X		X
Gall	X				
Van der Grinten		X	X		
Globular		X	X		

8.3 Conclusion

This text has attempted to bring together in one document a fairly complete set of map projection equations. To this end, it was necessary to define the problem and to introduce the terminology of mapping. Then, the basic transformations of mapping were derived. The model of the earth, with the current values of its parameters were then needed. Next, the three classes of map projections: equal area, conformal, and conventional, were considered one by one. For each class, a number of projection schemes were investigated. Finally, a means of numerically evaluating distortion was given. This has been an introduction to the theory of map projections.

However, of far more utility to the users of maps is the practice of map projections. Thus, in almost all cases, the final results of the direct transformation is expressed in planar cartesian mapping coordinates. These formulas for the most useful of the projecitons were made part of a computer program, and a variety of plotting tables were generated. The inverse transformations were given for a selection of the most important projections.

The level of mathematics chosen for the derivations was consistent with the strict intention of obtaining usable cartesian plotting equations.

Previous publications have suffered from one or the other of two opposite extremes. Many present the projection with little or no derivation, and intend to follow mainly a graphical approach. In these, the spheroid is not considered. The opposite approach is one of mathematical elegance. The same derivation is followed in a number of ways that seem to thrive on complexity. In this text the concepts of differential geometry were imposed as the unifying principles. In most cases, the derivations were relatively clean and straight-forward. In some other cases, such as the parabolic projection, it was necessary to use a unique method. Still, the derivations remained utilitarian.

The plotting tables serve in a number of ways. First, they were used to provide the data points for the graticules which appear as figures in Chapters 4, 5, and 6. Second, those plotting tables which are general, and depend on no standard parallel, can be used to produce a usable grid for any application. The user can choose his central meridian, and plot from there. Since scale depends on a multiplicative factor, a user can apply an engineer's scale to enlarge or contract the grid to his own specificaitons. Finally, the plotting tables, both those general and particular, can serve a user as a check if he prepares a mapping computer program.

The chapter on distortion gives a numerical means of estimating the amount of distortion in a particular area of a map. Again, the methods of differential geometry gave a unified approach to this in most cases. In the remaining cases, it was necessary to use a brute force approach to compare the length on the map to the length on a sphere or spheroid.

From the plotting tables, and the figures, it is apparent that at large scale, that is, over small areas, and in regions of minimal distortion, all of the projections approach the squares and rectangles of the Plate Carreé or the Carte Parallelogrammatique, except where meridian convergence is excessive. This is the reason that the grid system is useful on large scale maps.

Besides the use of natural and specialized projections, a number of variations are available by rotating the azimuthal plane, or the equatorial cylinder to cover areas of specific interest. Again, the plotting tables, and the subsequent figures indicate there are limited areas in each of these where distortion is minimized.

The reason that there are so many map projections is that there have been so many attempts to reach an acceptable representation of the spherical or spheroidal surface on a flat surface. Thus, map projections provide a variety of methods of transforming from an undulating earth to a flat piece of paper, and obtaining, in the end, a fairly reliable representation. It is the users duty to choose the map projection scheme that best satisfies his particular needs.

Bibliography

1. Adams, O. S., **General Theory of Polyconic Projections**, Special Publication 57, U.S. Coast and Geodetic Survey, 1934.

2. Adams, O. S., **General Theory of Equivalent Projections**, Special Publication 236, U.S. Cost and Geodetic Survey, 1945.

3. Adler, C. F., **Modern Geometry**, McGraw-Hill, 1958.

4. Bomford, G., **Geodesy**, Oxford, 1962.

5. Claire, C. N., **State Plane Coordinates by Automatic Data Processing**, Publication 62-4, Coast and Geodetic Survey, 1973.

6. Churchill, R. V., Brown, J. W., Verhey, R. F., **Complex Variables and Applications**, McGraw-Hill, 1960.

7. Davies, R. E., Foote, F. S., Kelly, J. E., **Surveying: Theory and Practice**, McGraw-Hill, 1966.

8. Deetz, C. H., and Adams, O. S., **Elements of Map Projections**, Special Publication 68, U.S. Coast and Geodetic Survey, 1944.

9. Deetz, C. H., **Cartography, a Review and Guide**, Special Publication 205, U.S. Coast and Geodetic Survey, 1962.

10. Goetz, A., **Introduction to Differential Geometry**, Addison-Wesley, 1970.

11. Grant, H. E., **Practical Descriptive Geometry**, McGraw-Hill, 1965.

12. Hershey, A. V., **The Plotting of Maps on a CRT Printer**, NWL Report No. 1844, June 1963, U.S. Naval Weapons Laboratory.

13. Hershey, A. V., **FORTRAN IV Programming for Cartography and Typography**, NWL Technical Report TR-2339, September 1969, U.S. Naval Weapons Laboratory.

14. Hildebrand, F. B., **Introduction to Numerical Analysis**, McGraw-Hill, 1956.

15. McBryde, F. W., and Thomas, P. D., **Equal Area Projections for World Statistical Maps**, Special Publication 245, U.S. Coast and Geodetic Survey, 1949.

16. Middlemiss, R. R., Marks, J. L., Smart, J. R., **Analytic Geometry**, McGraw-Hill, 1945.

17. Minor, D. C., and Denney, F. C., **College Geometry**, Prentice-Hall, 1972.

18. Moffitt, F. H., and Bouchard, H., **Surveying**, Intext, 1975.

19. Murdock, D. C., **Linear Algebra**, Wiley, 1970.

20. Pearson, F. F., **Map Projection Equations**, TR-3644, Naval Surface Weapons Center, 1977.

21. Pearson, F. F., **STAPLN: A Program for Direct and Inverse Transformation of State Plane Coordinates**, TN82-439, Naval Surface Weapons Center, 1982.

22. Richardus, P., and Adler, R. K., **Map Projections for Geodesists, Cartographers, and Geographers**, North Holland, 1972.

23. Seppelin, T. O., **"The Department of Defense World Geodetic System, 1972,"** The Canadian Surveyor, 28(5), 496-506, Dec. 1974.

24. Steers, J. A., **An Introduction to the Study of Map Projections**, University of London, 1962.

25. Thomas, P. D., **Conformal Projections in Geodesy and Cartography**, Special Publication 251, U.S. Coast and Geodetic Survey, 1952.

26. Wagner, K. H., **Kartographische Netzenwurfe**, Bibliographisches Institut Leipzig, January, 1972.

27. **CADDS-4, 2-D Mechanical Design Course Manual**, Edition No. 2, Computervision, 1982.

28. **SAS/GRAPH User's Guide**, Statistical Analysis Systems, 1981.

29. **Tables for a Polyconic Projection of Maps**, Special Publication 2, U.S. Coast and Geodetic Survey, 1946.

Subject Index